PEOPLE'S LIBER

Edited by
Susan M. Puska

August 2000

*****

The views expressed in this report are those of the authors and do not necessarily reflect the official policy or position of the Department of the Army, the Department of Defense, or the U.S. Government. This report is cleared for public release; distribution is unlimited.

*****

Comments pertaining to this report are invited and should be forwarded to: Director, Strategic Studies Institute, U.S. Army War College, 122 Forbes Ave., Carlisle, PA 17013-5244. Copies of this report may be obtained from the Publications and Production Office by calling commercial (717) 245-4133, FAX (717) 245-3820, or via the Internet at rummelr@awc.carlisle.army.mil

*****

Most 1993, 1994, and all later Strategic Studies Institute (SSI) monographs are available on the SSI Homepage for electronic dissemination. SSI's Homepage address is: http://carlisle-www.army.mil/usassi/welcome.htm

*****

The Strategic Studies Institute publishes a monthly e-mail newsletter to update the national security community on the research of our analysts, recent and forthcoming publications, and upcoming conferences sponsored by the Institute. Each newsletter also provides a strategic commentary by one of our research analysts. If you are interested in receiving this newsletter, please let us know by e-mail at outreach@awc.carlisle.army.mil or by calling (717) 245-3133.

ISBN 1-58487-027-3

CONTENTS

1. Introduction
   James R. Lilley 1
2. Going Places or Running in Place? China's Efforts to Leverage Advanced Technologies for Military Use
   Richard A. Bitzinger 9
3. PLA Logistics And Doctrine Reform, 1999-2009
   Lonnie Henley 55
4. Potential Applications of PLA Information Warfare Capabilities to Critical Infrastructures
   William C. Triplett II 79
5. China's Military Space and Conventional Theater Missile Development: Implications for Security in the Taiwan Strait
   Mark A. Stokes 107
6. PLA Air Force Operations and Modernization
   Kenneth W. Allen 189
7. The Kosovo War: Implications for Taiwan
   Arthur C. Waldron 255
8. China's Maritime Strategy
   Bernard D. Cole 279

About the Authors 329

Index 333

CHAPTER 1

INTRODUCTION

James R. Lilley

An analytical schism has developed over differing assessments of China's military modernization. Underlying this debate are at least two key questions. First, will the ongoing China's People's Liberation Army (PLA) modernization provide China with significant offensive power projection and/or preemptive capability? If so, by when? Second, does the pace and success of China's military modernization constitute a threat to the United States and/or its friends and allies in the Asia-Pacific region?

At the source of these differing views on the pace and likely success of the PLA modernization is a lack of hard evidence, aggravated by a Chinese tendency to conceal both strengths and weaknesses. There are also analysts who are locked into positions on the PLA that the evidence seems unable to alter. Lack of information is often muddied by anecdotal knowledge, sometimes provided by Chinese interlocutors, that may be impossible to confirm or refute. In addition, a large body of conventional wisdom about the PLA has built up over time, which may inhibit fresh reassessment. Finally, peer comparisons of the PLA to the U.S. military, which is without equal in the post-Cold War period, may shape analysis of the PLA's capabilities and shortfalls. As a result, conclusions about China's military modernization often leave considerable room for interpretation on any side of an argument.

The policy decisions made today based in part on the absence of hard analysis will likely haunt U.S. and allied policy and interests well into the 21st century, as China's comprehensive strength and historic aspirations mature. In order to minimize miscalculations about the PLA modernization, debate is essential, for it can help reduce the twin analytical evils of overestimation and underestimation. Debate can reduce wishful thinking or demonization about China's strategic capability and intentions. It may also preclude counterproductive self-imposed constraints on U.S. policy options based on fear of self-fulfilling prophecies of a China threat. Active debate, finally, can lead to a reevaluation of old, well-worn assumptions, and spur greater exploitation of publicly available information about the PLA and China's national security. All of this would potentially help to develop a clearer picture of China's People's Liberation Army After Next into the 21st century.

The 1999 PLA Conference, which was hosted jointly by the American Enterprise Institute and the U.S. Army War College's Strategic Studies Institute, convened September 10-12, 1999, at Carlisle Barracks, Pennsylvania. The goal of this conference was to comprehensively examine Chinese military modernization efforts. The meeting drew together leading experts on the PLA army, navy, air force, missile forces, and national defense industries and included PLA experts with opposing views on the pace and likely success of Chinese military modernization. Lively debate continually probed analytical differences and prejudices, as well as the sources of information upon which conclusions were based. The conference also included a preliminary yet timely examination of the PLA's potential application of information warfare. An initial discussion of the post-Kosovo implications for China's Taiwan strategy and China's foreign military relations also took place.

"Going Places or Running in Place? China's Efforts to Leverage Advanced Technologies for Military Use," by Richard Bitzinger, opened the conference, and is Chapter 2 in this volume. Bitzinger finds that China's military-industrial complex has achieved some success in technology leveraging over the past 20 years, and as a result, it has improved its production capabilities to develop and deploy some relatively modern weapon systems. In particular, it has made real progress in the areas of ballistic and cruise missiles.

Overall, however, these accomplishments tend to be the exceptions that prove the rule. Particularly when it comes to fighter aircraft, surface combatants, and ground equipment, the Chinese still confront severe problems when moving from prototypes to production, including drawn-out development times, program slippage, and small and fitful production runs. Bitzinger argues that China's defense industries fail mainly due to endemic technical, structural, and cultural problems. A shortage of technical skills, overcapacity, underfunding, and a bureaucratic, hierarchical and risk-averse work environment have all impeded technology leveraging. Moreover, he contends that current defense industry reforms do little to remove or overcome these impediments. As a result, he concludes the PLA will find it difficult to rely on its domestic defense industry to meet many of its more ambitious, near-term military modernization goals.

Chapter 3, "PLA Logistics and Doctrine Reform, 1999-2009," by Lonnie Henley, discusses operational changes, as well as the 10-year effort announced by General Wang Ke, Chief of the General Logistics Department, in November 1998, to restructure and reform PLA logistics to support mobile warfare. Henley argues that logistics reforms address critical weaknesses in the PLA's ability to fight a modern war in or near Chinese territory, while operational changes seek to combine restructuring of ground forces with improved doctrine. Together these changes will help the PLA to "progress toward competent combined-arms operations" with effective joint operations as the ultimate goal. Henley assesses that these changes, even if successful, may not significantly increase China's power projection capability. He does argue, however, that changes in logistics and operations will improve the PLA's ability to move and sustain its forces within China and around its periphery.

In Chapter 4, "Potential Applications of PLA Information Warfare Capabilities to Critical Infrastructures," William C. Triplett II discusses the potential for the PLA to conduct offensive information warfare operations against critical civilian infrastructures in the United States, and how such an operation might unfold. Triplett's discussion is particularly timely in the wake of the May 2000 "I Love You" software virus that crippled hundreds of thousands of computer systems throughout the world,[1] reportedly penetrating even into classified Pentagon systems.[2] Triplett notes the PLA has shown intense interest in information warfare with top level support of PLA leaders, such as General Fu Quanyou, Chief of the General Staff, PLA, and generous funding. He observes that PLA Informational Warfare theorists stress the need for a preemptive strike capability, which Stokes and others in this volume note has become a fundamental underpinning of the PLA's overall military strategy of "active defense." Triplett argues that the United States, Taiwan, and Japan (a key ally in the Asia-Pacific region) are vulnerable to cyber-attack on critical transportation, communications, and financial networks. Triplett concludes with a warning that the "gap between what information warfare is capable of accomplishing and the non-action by the defenders" is growing daily, which demands greater attention and action in the United States, Japan, and Taiwan.

"China's Military Space and Conventional Theater Missile Development: Implications for Security in the Taiwan Strait," by Mark A. Stokes is Chapter 5. Stokes asserts that China is establishing "one of the most daunting conventional theater missile challenges in the world," which the PLA views as an asymmetrical "trump card" against a superior force. He argues that theater conventional missiles, supported by spaced-based reconnaissance and combined with a preemptive strategy, provide Beijing with a keen psychological tool and possible military advantage, particularly in the Taiwan Strait. China's theater conventional missiles and space reconnaissance architecture are emerging as the cornerstone of Chinese warfighting strategy. They form a relatively cheap and expeditious compensation for shortcomings in China's navy and air force (see Chapters 6 and 8 in this volume), and they are being developed based on the PLA's assessment of U.S. military weaknesses—reliance on space systems, aircraft carrier battle groups, and expeditionary air forces.

Stokes also asserts that a disturbing convergence has occurred between Chinese leaders and American political and sympathetic academic circles who argue that any missile defense against the PLA's growing arsenal of increasingly accurate and lethal missiles would be destabilizing and lead to an arms race. Stokes counters this argument by noting that, if the PLA achieves an overwhelming offensive missile advantage, this would intensify the already existing arms race and could destabilize the cross strait balance, causing Taiwan to "shift toward a tactically offensive doctrine," including a nuclear weapon, to deter the PLA's overwhelming missile threat.

In Chapter 6, "PLA Air Force Operations and Modernization," Kenneth W. Allen argues that the PLA Air Force (PLAAF) is undergoing a "crucial transition" from an obsolete giant with short operational legs and limited all-weather capability to an "offensive-oriented force with extended range and greater lethality," able to fight local, limited wars under high-tech conditions. To achieve this transformation, the PLAAF has expanded its defensive doctrine to include both offensive and defensive operations. Allen points out that with the modernization of its equipment, the PLAAF has been able to modernize its tactics, extend its combat range, and practice providing support to naval and ground operations. In addition, Allen describes the transformation of the 15th Airborne Army into a formidable rapid reaction force with three divisions as one of the most important changes in recent PLAAF campaign strategy.

Despite improvements across the board, however, Allen argues that the PLAAF still lags far behind its modernization goals. It retains a mixed fleet of modern aircraft, such as the J-11 (SU-27), alongside 1950s vintage aircraft, while its ability to conduct all-weather flying and extended operations—essential to support an offensive strategy—are questionable. Further, China's ability to exploit foreign advanced technologies and equipment in production and utilization of advanced aircraft is (as Bitzinger has also observed) a limitation. Still, Allen concludes that PLAAF shortfalls will not inhibit its dedication to completing its assigned missions, nor do its shortfalls reassure regional powers of China's long-term regional capabilities and intentions. As Allen points out, based on recent research throughout the region, Asian countries watch the PLA very closely and with growing concern that it could develop sustained combat capabilities within the next 10-15 years.

In Chapter 7, "The Kosovo War: Implications for Taiwan," Arthur C. Waldron addresses the lessons of Operation ALLIED FORCE on China's calculus to use force against Taiwan, if necessary, to quickly and decisively impose reunification as a *fait accompli* before the United States has time to intervene. Beneath China's frantic rhetoric, which was feverishly fanned by the accidental bombing of the Chinese Embassy in Belgrade, is the recognition that time is running out for China. The concept of sovereignty is less sacrosanct now among the international community, which has demonstrated that it will act to stop genocide and other atrocities within a country's borders. Waldron asserts Operation ALLIED FORCE also demonstrated the limitations of military force to resolve issues. Its lessons suggest that the application of force is no longer a viable option for China to resolve the Taiwan problem. Further, Taiwan's indigenous democracy increasingly is recognized by the international community as legitimate, which makes the possibility of the use of force even more complicated and risky for China.

Chapter 8, "China's Maritime Strategy," by Bernard D. Cole, examines the development of China's maritime strategic thinking and its recent attempts to transform the PLA Navy (PLAN) from a coastal, "brown water" (out to 100 nautical miles) to an open ocean "blue water" force that is capable of reaching the "second island chain," defined as a line from Japan through the Bonin, Marianas, and Caroline Islands by 2020. The PLAN faces daunting limitations in its efforts to develop into a power projection force, according to Cole. The most serious shortfall for the PLAN is a lack of air power at sea, but it also lacks in virtually all categories affecting modern naval operations—training and education; naval systems and platforms; naval warfare technology and systems; doctrine and tactics; command and control; intelligence; strategic planning; leadership; and influence within the national strategy-making structure. Cole observes that the PLAN's stated goal to achieve a "blue water" navy, which presently targets U.S. naval capability, serves as an impetus to protect and expand the PLAN's share of the military budget. Cole notes that, despite recent purchases from the Russians and overall improvements of the navy, there is no evidence to support that the PLAN has initiated the major modernization effort that would be required to actually transform the largely coastal navy into a power projection force by 2020.

In summary, Cole and Allen raise important questions and considerable doubt about the likely success of PLAN and PLAAF modernization, respectively, over the next 10-20 years. Bitzinger also seriously questions the efficacy of China's national defense industries to sustain modernization, while Henley raises skepticism about whether China's military logistics support, an essential underpinning of power projection, can reach its goals.

Nonetheless, all agree that China actively seeks a strong, modern military force, with power projection capability to protect and promote China's national interests within the region and beyond, and that it has openly added "preemptive strike" to its operational strategy. Since China clearly recognizes its own shortcomings, and the gap between its military capabilities and those of the United States, as well as against a regional power such as Japan, it seeks ways to compensate for these deficiencies with surprise and asymmetrical operations, perhaps including information warfare and conventional missile strikes. The PLA's ongoing military modernization and emphasis on surprise and preemption combine with China's refusal to renounce the use of force against a democratic Taiwan and its territorial claims over the South China Sea to heighten concern about China's intentions and capabilities within the region.

CHAPTER 1 - ENDNOTES

1. John Markoff, "A Rogue Software Program Attacks Computers Worldwide," *New York Times*, May 5, 2000, p. 1.

2. Associated Press, "Virus Infects Classified Pentagon Systems," *Colorado Springs Gazette*, May 6, 2000, Early Bird, May 8, 2000.

CHAPTER 2

GOING PLACES OR RUNNING IN PLACE? CHINA'S EFFORTS TO LEVERAGE ADVANCED TECHNOLOGIES FOR MILITARY USE

Richard A. Bitzinger

China's current efforts to modernize its armed forces have been the subject of considerable foreign inquiry and assessment.[1] As part of this modernization process, Beijing has gone to considerable lengths to find and acquire advanced defense and defense-related (i.e., dual-use) technologies—both through indigenous research and development (R&D) activities and via foreign inputs. Beyond the mere **acquisition** of potentially useful technologies, however, China must also be able to effectively **leverage**—that is, to absorb, assimilate, and exploit—these technologies for military purposes.

This chapter specifically addresses China's recent efforts to leverage advanced technologies for military use. The conclusion reached here is that the Chinese military-industrial complex continues to experience considerable difficulties in assimilating and exploiting advanced technologies for military purposes, particularly imported technologies. Admittedly, China's military-industrial complex has experienced some genuine successes in technology leveraging over the past 20 years, and as a result, it has made incremental improvements both in its military products and production capabilities. Beijing is certainly building some better weapons, and its defense factories are considerably improved. On the whole, however, Chinese success with leveraging advanced technologies has been—and will likely remain—narrowly constrained, fitful, and slow. In some areas, China's military-industrial complex has extracted a few tangible benefits from advanced technologies inputs, while in other areas developments have been insufficient to close critical technological gaps with likely military competitors. Moreover, this chapter finds that these failures are due mainly to technical, structural, and cultural impediments inherent in the country's defense industrial base, and that current efforts to reform China's defense industry are doing little to remove or overcome these impediments. As a result, Beijing will continue to find it difficult to rely predominantly on its indigenous defense industry in order to acquire advanced conventional weapons necessary to meet its military modernization goals.

At the same time, it is acknowledged here that various gaps and limitations exist in this chapter. By focusing mainly on the "kinetic" aspects of military power—that is, lethal items like missiles, combat aircraft, warships, and ground ordnance, along with their main subsystems and components (like radar and defense electronics)—one is obviously omitting much of the field of information-based systems, an admittedly embryonic area of warfighting where the Chinese may be making more progress in technology assimilation and application than is immediately evident. This is an area that demands closer investigation. In addition, there are limits to empirically analyzing and assessing Chinese success and failure with technology absorption and leveraging, at least where the military-industry complex is concerned. One is bounded by what evidence one can find in the public record, which is often less than satisfactory in terms of detail or amounts of information. In addition, what one sees is open to differing interpretations. Finally, much of this information is secondhand and filtered through someone else's perceptions, creating even more subjectivity. Despite these drawbacks and limitations, the inductive approach used here can still be useful in providing insights into China's experiences with leveraging technologies for military purposes.

Chinese Technology Acquisition Activities.

Several Western analysts have meticulously documented Chinese efforts to develop advanced technologies for actual or projected military uses. These writings indicate a particular interest on the part of the Chinese military in such technology areas as long-range precision-strike, command, control, communications, computing, and intelligence (C^4I), information warfare, and area denial.[2] For example, China is working on many of the requisite technologies necessary to producing an indigenous land-attack cruise missile (LACM), including airframe design (perhaps adapting its C-802 antiship cruise missile to a LACM profile), propulsion systems (such as small turbojet engines and ramjets/scramjets), and guidance technologies (such as GPS for in-flight navigation and terrain contour matching guidance[TERCOM], imaging infrared, or synthetic aperture radar [SAR] for terminal homing).[3] In addition, it is developing laser-, IR-, or TV-guided precision-guided munitions, railgun and microwave weapons, and antisatellite (ASATs) laser weapons, global positioning system (GPS) guidance for ballistic missiles, multiple rocket launchers armed with smart submunitions, sea mines, and a seeker for an active-radar-homing air-to-air missile.[4] China has also demonstrated at least some low observability (LO) technological capabilities for reduced radar cross-sections and heat signatures. At the 1998 China Airshow in Zhuhai, for example, a Chinese company, Seek Optics, claimed that it had developed a number of radar-absorbent materials and coatings, which could be applied to aircraft, cruise missiles, or warships.[5]

Much of China's recent technology development activities are civilian in nature and in origin, albeit with considerable potential for commercial-to-military spin-on.

Beijing is conducting a wide variety of basic and applied R&D, designed to raise the general level of the country's science and technology (S&T) base, including solid-state phased-array radar, ultra-wide band radar technology and photoreconnaissance, and remote sensing.[6] The military also stands to gain by piggybacking on technology breakthroughs initiated via the country's "863" and "Super-863" programs. The 863 Program was initiated in March 1986 as an essentially civilian/commercial S&T development program, but with considerable potential for military spin-on—signified by the fact that it was co-managed by the Commission on Science, Technology, and Industry for National Defense (COSTIND), China's main defense R&D oversight organization. The 863 Program concentrated on seven main areas for long-term S&T development: astronautics, information technologies, biotechnology, lasers, automation, energy, and new materials. The Super-863 Program, which succeeded the 863 Program in 1996, focused subsequent long-term Chinese S&T development in such key areas as machine tools and computerized manufacturing systems, microelectronics and telecommunications, bioengineering, exotic materials, and nuclear, aviation, space, and marine technologies.[7] For example, China is working on forming a national information infrastructure for improved C^4I interconnectivity, including a high-speed fiber-optic computer network and a satellite communications system.[8]

In addition to indigenous S&T developments, China has greatly increased its imports of military technology and know-how. According to research done by Richard Fisher,[9] Bates Gill and Taeho Kim,[10] and others, China has acquired a broad array of defense technologies from foreign sources—primarily Russia, but also Israel and Western Europe—including:

- **Cruise missiles.** Several Russian institutes reportedly have sold cruise missile technology and have provided Russian advisors and technicians to work on Chinese cruise missile programs. For example, China may be receiving Russian stealth technology applicable to fielding a low-observable cruise missile. In addition, Israel reportedly has supplied China with technology relating to a jet-powered drone, which could conceivably be adapted to attack.[11]

- **Reconnaissance satellites.** China and Brazil plan to orbit a series of earth observation satellites, called CBERS; the first CBERS satellite will likely be operational by 2000. At the same time, Canada and Russia reportedly are providing China with SAR sensor technologies for earth-observation satellites that could also be applied to military imaging satellites.[12]

- **Electronic intelligence (ELINT).** Israel and Russia reportedly have supplied ELINT equipment to the People's Liberation Army (PLA), which is being incorporated into special, missionized aircraft.[13]

- **Medium-range surface-to-air missiles (SAMs).** China reportedly has received *Patriot* SAM technology via Israel, as well as acquiring several hundred SA-10 SAMs from Russia; according to Fisher and others, technology from these missiles could be exploited for China's indigenous HQ-9 medium-range SAM program.[14]

- **Fighter aircraft.** China is currently producing under license up to 200 Russian-designed fourth-generation Su-27 fighters. In addition, China's developmental J-10 (F-10) fighter reportedly is heavily based on technology derived from Israel's cancelled Lavi fighter jet program.[15]

- **Combat radar and other avionics.** Russia, Israel, and Western Europe are providing radar and avionics to a number of Chinese development combat aircraft programs—including the FC-1 and the J-8IIM fighters. Although most of these programs are for re-export, such cooperation may provide China with opportunities for technology exploitation that could advance China's own indigenous R&D bases.[16]

- **Air-to-air missiles (AAMs) and precision-guided munitions (PGMs).** Israel reportedly has granted China a license to produce its short-range *Python-3* AAM. In addition, China has reportedly purchased laser-guided artillery projectiles and bombs, and TV-guided PGMs from Russia.[17]

- **Submarines.** The Russian submarine-design bureau Rubin reportedly is assisting China in developing its next-generation nuclear attack submarine (SSN), which could also aid Chinese efforts to build a new nuclear-powered, ballistic-missile-carrying submarine (SSBN).[18]

In addition to providing direct technology assistance to China's military-industrial complex, foreign suppliers may be inadvertently advancing Beijing's military modernization efforts by selling it equipment that the Chinese could either reverse-engineer or use for the purposes of technology exploitation. The Chinese have already reverse-engineered several pieces of military equipment, including France's AS-350 helicopter (produced by China as the Z-11[19]), the Italian *Aspide* AAM/SAM (built by China as the PL-11 AAM and the LY-60 SAM[20]) and the French *Crotale* short-range surface-to-air missile (produced by China as the FM-80). The Chinese PL-9 short-range air-to-air missile reportedly is a reverse-engineered version of the Israeli-supplied *Python-3* AAM.[21] In addition, according to the Cox Report, in the early 1980s the Chinese purchased two CFM-56 turbofan engines (used in commercial and military transport aircraft) and attempted to reverse-engineer the engine, ostensibly for military uses.

In particular, foreign **commercial** technology transfers—particularly those implemented under the auspices of defense conversion—have provided Beijing with opportunities to acquire advanced dual-use technologies that could be redirected to the arms industry. The West has been a critical supplier of investments and technologies that are helping China develop civilian high-tech sectors within its defense industry, which in turn could help underwrite the design and manufacture of sophisticated weapon systems. Such dual-use technology transfers include not only "hardware"—such as technical specifications and licenses (e.g., for the local production of advanced materials, metallurgy, propulsion, computers, microelectronics, and electro-optics) and production and process technology (such as sophisticated machine tools and workforce training), and, above all, money (to support the modernization of Chinese factories and product lines)—but also "software"—more intangible but nevertheless critical elements such as quality assurance and Western management and marketing skills, which could indirectly aid defense production.[22]

For nearly 20 years, Western aerospace companies have been involved in extensive technology transfers to China's commercial aviation industry. For example, the McDonnell Douglas Corporation (now part of the Boeing Aircraft Company), as part of its so-called Trunkliner program with China, established a production and assembly line in Shanghai to build its series of MD-80 and MD-90 passenger jets. As part of this arrangement, the Chinese purchased from McDonnell Douglas a number of large computerized, numerically controlled machine tools, including multi-axis milling and profiling machines, that the company had lying unused in a closed aircraft plant.[23] Boeing and the European Airbus consortium have also helped establish production facilities in China for subassemblies and parts production. In addition, Eurocopter, Sikorsky Helicopter, Pratt & Whitney jet engines, Bombardier of Canada, Hexcell, ATR (a Franco-Italian regional aircraft consortium), and Allied Signal have all established joint ventures in China for coproducing aerospace systems or components.[24] The Cox Report also asserts that Beijing attempted to use a joint venture with Pratt & Whitney, Canada, to help it develop a military jet engine.[25]

China's shipbuilding industry has also benefited greatly from both commercialization and foreign technology inputs. According to Evan Medeiros, beginning in the 1980s,

Chinese shipyards successfully converted much of their production to profitable civilian products, such as bulk carriers and general cargo ships. Moving into commercial shipbuilding both demanded and permitted the extensive modernization and expansion of Chinese shipyards; over the past 15-20 years, these yards have added greatly to their productive capacity, building huge dry-docks and heavy-lift cranes. At the same time, they entered into a number of technical cooperation agreements and joint ventures with shipbuilding firms in Japan, South Korea, Germany, and other countries, providing the Chinese with advanced shipbuilding design and manufacturing technologies—in particular, computer-assisted design and manufacturing, hull construction integration systems, propulsion systems, and numerically controlled processing and testing equipment—training, capital, and other know-how. Medeiros adds that commercialization and internationalization has led to important changes in the way Chinese shipbuilding enterprises are run, resulting in a more market-oriented and decentralized management. As a result, military shipbuilding programs collocated at these yards theoretically should be able to take advantage of these infrastructure and software improvements to design, develop, and construct improved warships. Some Chinese shipbuilding facilities can build ships of 150,000-200,000 tons—large enough to permit construction of an aircraft carrier.[26]

Another critical civil/commercial technology area with far-reaching implications for spin-on is microelectronics. The development of a world-class microelectronics sector has been a government industrial development since the early 1990s.[27] Particular attention has been paid to creating an indigenous semiconductor manufacturing capability. The most significant manifestation of this effort has been the 909 Project, which entailed the establishment of the $1.2 billion Huahong Company near Shanghai—built with technology and funding provided by Japan's NEC Corporation—which this year began production of

64-megabit memory chips with 0.30 micron widths. Huahong-NEC is currently turning out 5,000 200mm (8-inch) wafers a month, and production will eventually rise to 20,000 wafers a month.[28]

The United States has also been an important supplier of dual-use information technologies that could be redirected towards China's military-industrial complex, including computers, encryption technology, and fiber-optic and microprocessor manufacturing equipment.[29] Thanks to these and other foreign investments, China is becoming increasingly proficient in the areas of telecommunications, semiconductors, software, and information-processing—all of which provide China with growing opportunities for spin-on, particularly when it comes to developing tools for harnessing the information revolution in warfare.

Technology Leveraging: An Empirical Assessment.

China has clearly gained access to considerable amounts of militarily useful technologies and know-how. The next step is to assess how well China's defense industry is doing in effectively leveraging (i.e., absorbing and exploiting) these technologies. In this regard, at least two empirical approaches are possible. First, one can attempt to infer Chinese progress in technology absorption and leveraging by examining its success in getting positive results from its military-industrial complex in the form of new advanced weapon systems. In particular, such a "results-based" approach should look not only at what the Chinese are capable of developing, but also their progress in translating R&D developments into the timely, serial production of sophisticated weaponry. Second, at the level of armaments production, there is a growing body of anecdotal evidence to permit at least a partial assessment of Chinese capabilities in leveraging advanced manufacturing technologies, particularly in the aviation and shipbuilding industries.

Technology Leveraging at the Product Level: A "Results-Based" Analysis. China, of course, has made its most significant production breakthroughs in the area of missile systems, especially surface-to-surface ballistic missiles (SSMs) and antiship cruise missiles (ASCMs). In fact, it has become almost *de rigueur* to refer to China's missile industry as an "island [or pocket] of excellence" in the country's military-industrial complex. Beijing has indigenously developed a full range of surface-to-surface missiles—from short-range artillery rockets to medium-range, road-mobile systems to intercontinental and submarine-launched ballistic missiles (ICBMs and SLBMs). Moreover, China has demonstrated the ability to produce solid-fueled and/or multistage missile systems, including space-launch vehicles.

With regard to tactical missile systems, the 600 km-range DF-15 (M-9) and the 300 km-range DF-11 (M-11) SSMs are perhaps the most well-known. According to Wendy Frieman, the DF-15 and DF-11 missiles "represent[s] another class of weapon that China did not have in the early 1980s—solid-fuel rocket motors—**largely based on indigenous R&D**."[30] Moreover, China is expected to incorporate satellite-assisted navigation (probably GPS or the Russian GLONASS system) into these missiles to improve their accuracy.[31] Production of these two missile systems appears to be already well-underway: Press reports indicate that the number of DF-15 and DF-11 SSMs deployed along Taiwan Strait has grown from 30-50 around the time of the 1995/1996 Straits crisis to 160-200 in early 1999, and that this figure could rise to 650 by 2005—indicating the Chinese could be producing around 50-100 of these missiles annually.[32]

Another bright spot in the country's military-industrial complex is the antiship cruise missile sector. Again, foreign technology inputs have been critical to Chinese weapons development. The Silkworm family of ASCMs, of course, is based on the Soviet SS-N-2 *Styx*, while the more modern C-801 and C-802 ASCMs are reportedly derived from the French *Exocet*.[33] Moreover, France reportedly supplied a small turbojet engine to power the C-802.[34] More recently,

China has unveiled an entirely indigenous and all-new ASCM: the short-range, TV-guided C-701, displayed publicly for the first time at the 1998 China Airshow in Zhuhai.[35] China's success with developing and producing reasonably advanced ASCMs suggests "an ability to integrate relatively modern Western technology into existing Chinese designs."[36]

Directly drawing upon its successes in producing antiship cruise missiles, China is reportedly progressing on the development of a land-attack cruise missile. A January 2000 article in *Jane's Defence Weekly* asserts that China's main ASCM academies have been working on a LACM—known as the X-600 or HN-1—since the late 1970s.[37] The missile is similar in appearance to the Russian Kh-55/AS-15 strategic cruise missile, although with a shorter range (600 kilometers), fueling speculation that the X-600 could be a derivative of the Kh-65 missile (itself a short-range version of the Kh-55), which Russia has reportedly transferred to China.[38] *Jane's* states that the X-600/HN-1 entered service with the PLA in 1992, but this has not been confirmed officially. At the very least, however, China could deploy a rudimentary land-attack cruise missile by the middle of the next decade.[39]

Other areas of missile development and production where China is apparently demonstrating some success in exploiting imported technologies are air-to-air missiles—such as the PL-9 (a modified *Python-3*) and the PL-7 (believed to be a reverse-engineered version the French-designed R550 *Magic* AAM[40])—and surface-to-air systems—such as the QW-1 shoulder-launched SAM (which reportedly incorporates *Stinger* technology[41]), and the FM-90 SAM (an upgraded version of *Crotale*-derived FM-80, featuring a faster missile and a longer range[42]).

A few other sectors outside of the missile industry also show promise. One of these is the diesel submarine (SSK) shipbuilding industry: China has developed a modern indigenous SSK—the *Song*-class—featuring a more hydrodynamically efficient design and an asymmetrical seven-bladed propeller, for quieter running.[43]

With regard to other programs, however—especially combat aircraft, surface combatants, and ground equipment—the Chinese would still appear to confront severe problems when it comes to moving prototypes into production. These difficulties, in turn, would suggest continuing problems with mastering the myriad technologies that go into such military equipment. One indicator is the long development times and program slippage that many weapons experience before finally entering production—a problem the Chinese readily concede.[44] The JH-7 fighter-bomber, for example, was initiated over 20 years ago and first flew in the late 1980s; however, it did not enter even low-rate production until 1997 or 1998.[45] The J-10 fighter did not fly until early 1998—nearly 15 years after its program start.[46] According to open sources, the J-10 is unlikely to enter production before the middle of the next decade—approximately 25 years after the program began![47]

After the Chinese begin building a weapon system, production runs are often small and fitful. According to estimates made by PLAAF analyst Ken Allen, the Chinese are probably manufacturing only around 36 fighter aircraft a year—12 J-8IIs and 24 J-7s.[48] In addition, although Beijing acquired a license in the early 1980s to produce the French-designed AS-365N utility helicopter (called the Z-9 in China), the Chinese have reportedly built no more than 50 to 80 Z-9s over the past two decades—and in sporadic batches, at that.[49] With regard to naval shipbuilding, since 1991 China has launched only three destroyers and nine frigates—barely 1.5 major surface combatants per year.[50] Production of the *Song*-class submarine has been equally sporadic, with only one boat commissioned so far (in 1998, after being launched in 1994). As a result, the Chinese continue to produce the near-obsolete *Ming*-class submarine (a copy of the Soviet Romeo design, dating from the late 1950s).[51]

These program stretchouts and small production runs are all the more significant given that much of what the Chinese are currently building is not particularly cutting-edge. The J-7 (MiG-21) fighter is essentially late 1960s/early 1970s technology, while the J-8II is roughly equivalent to a late-generation F-4 fighter (approximately late 1970s technology). China's 20-year teething problems with the JH-7 fighter-bomber involve an aircraft of rather unremarkable design—i.e., no fly-by-wire flight control system or exotic materials in its construction. Even the J-10—admittedly, a full-up fourth-generation fighter—is basically 1980s technology that will not be deployed until around 2005.[52] For its part, the first *Song*-class diesel submarine has been described as a "patchwork of systems and technologies," "very noisy with a lot of equipment problems," and "effectively shown to be a failure."[53]

Even the defense industry's "island of excellence" has shown some less-than-impressive results. Many of China's newest missile systems remain at least one or two generations behind that of the West—basically comparable to 1960s- or 1970s-era technology—such as the HJ-8 antitank guided weapon (reportedly derived from the Soviet-designed AT-4 *Spigot*), the PL-11 semi-active radar-homing AAM (based on the Italian *Aspide*), and even the C-802 ASCM (roughly equivalent to the 1970s-era *Exocet*).[54] And even if China were to field a fully active radar-homing (ARH) AAM by 2005, it would still be nearly 25 years after the United States first began production of the AMRAAM missile; moreover, most Western nations are already working on the next generation of ARH AAMs, employing ramjet propulsion and improved seeker technologies. Finally, it says much about the likely poor capabilities of two highly touted (and widely marketed) Chinese ASCMs—the supersonic C-101 and C-301 missiles—that no military—not even the PLA—has purchased these weapons.

Oft times, too, "progress" turns out to be superficial or illusory. While China is clearly advancing its space-based capabilities for reconnaissance, communication, and navigation, it is still years away from acquiring even a rudimentary military capability. The first China-Brazil Earth Reconnaissance Satellite (CBERS), launched in October 1999, has only a 20-meter resolution (compared to one meter or less found on most military reconnaissance satellites), while it and other planned Chinese earth-observation satellites will continue to lack real-time surveillance or quick revisit capabilities. Meanwhile, China's efforts to expand its microelectronics industry continue to suffer setbacks. In particular, semiconductors produced at its showcase Project 909/Huahong-NEC factory in Shanghai experienced a 50 percent failure rate in its first year of operation, and sales have been poor.[55]

Moreover, despite years of arduous R&D efforts, China's defense industry continues to rely heavily—and perhaps increasingly—upon foreign technologies. These dependencies are especially acute when it comes to jet engines, marine diesel engines, and fire-control radar and other avionics. The J-10 fighter, for example, is reportedly powered by the Russian-built AL-31F engine, which is also used in the Su-27.[56] At the same time, endemic "technical difficulties" surrounding the JH-7 fighter-bomber's indigenous engine have resulted in significant program delays and ultimately forced the Chinese to approach the British about buying additional *Speys* in order to continue aircraft production.[57] The new *Song*-class submarine uses a German-supplied diesel engine,[58] while both the *Ming*- and *Han*-class submarines reportedly have been upgraded with a French sonar and combat system.[59] China's new *Luhai*-class destroyer incorporates a number of foreign-supplied systems, including a Ukrainian gas turbine engine, a German electrical system, Italian torpedoes, and Russian helicopters.[60] Finally, the Chinese have yet to develop an indigenous turbine engine transmission (considered only "mid-level" technology) for its armored vehicles; consequently, they have turned to the Europeans for help.[61]

To an extent, it makes sense for Beijing to import advanced technologies, rather than waste time and resources duplicating the same capabilities indigenously. In addition, by integrating these technologies into domestic systems, the Chinese are demonstrating some capability to leverage foreign technologies. Nevertheless, a growing dependency on imported systems and technologies is also a strong indicator of likely deficiencies within the country's military-industrial complex. For example, China's current export-oriented fighter programs—the FC-1 and the J-8IIM—respectively incorporate a West European or Russian radar and avionics suite; in addition, the FC-1 is powered by a Russian engine. Finally, the PLA's penchant for off-the-shelf arms imports could be interpreted as an indicator of growing frustration on the part of the Chinese military with the defense industry's inability or delay in development and producing advanced weapon systems. The PLA, for example, has recently concluded a number of foreign weapons buys, including purchases of Su-30 fighter-bombers, medium-range SA-15 SAMs, *Sovremennyy*-class destroyers, and Israeli AWACs systems, in addition to the recent startup of Su-27 licensed-production.[62]

Reverse-engineering has its own limitations: While the Chinese often have been able to "indigenize" several types of foreign missile systems, they have had much less success with fighter aircraft (such as the MiG-23[63]), helicopters (such as the French SA-321G *Super Frelon* helicopter [Z-8][64]), or jet engines (such as the *Spey*, which powers the new JH-7 fighter-bomber[65]). In other cases, the Chinese must turn to the original manufacturer for critical components in order to successfully produce the copied system. Larry Wortzel, for example, points out that Beijing reverse-engineered the U.S.-designed AN/TPQ-37 artillery-locating radar, which China produces as the Type-704; at the same time, the Chinese have attempted to order a large number of spare parts for the AN/TPQ-37, indicating an inability to copy these components.[66] Even when successful, reverse engineering is often a time- and resource-consuming chore—sometimes to the point of starving other, more cutting-edge R&D. As a result, by the time the Chinese have perfected a reverse-engineered system, the "state-of-the-art" has progressed to the next level, leaving the PLA with a weapon that, although an improvement to its current arsenal, does little to narrow the technological gap with its competitors.[67]

Technology Leveraging at the Level of Arms Production: The Anecdotal Evidence. Besides problems with technology leveraging at the product level, large parts of China's arms industry are experiencing frustrations with leveraging advanced manufacturing technologies **at the level of armaments production**. In the case of the aviation industry, deficiencies in aircraft design capabilities, metallurgy, avionics, and engine technology "have been common and have prevented mass production."[68] According to a U.S. aviation industry representative, Chinese aircraft factories were "less than optimum for the task" of cooperative manufacturing.[69] Jonathan Pollack and James Mulvenon argue that the Chinese aviation industry, while able to perform most of the tasks necessary to the assembly of large commercial aircraft, was still deficient in several areas of aircraft design, development, and production—and while their assessments were largely directed at China's **civil** aircraft sector, these points appear applicable to China's advanced military aircraft programs as well.[70]

The Chinese are already experiencing problems and delays with their Su-27 co-production program—even though it currently entails the relatively simple task of assembling knock-down kits imported from Russia. While the first two locally assembled Su-27 fighters achieved first flight in December 1998, press reports indicate that the aircraft immediately had to be taken apart and rebuilt due to "sub-standard work." This same report states that, due to problems with absorbing advanced manufacturing technologies and processes, it will be at least 2 years before the Chinese aircraft factory building the Su-27 will reach full-rate production; as a result, China will be able to produce only six or seven Su-27s annually over the next 3 years.[71] Consequently, at least 100 Russians will remain on-site at the Su-27 plant to oversee production and guarantee quality control.[72]

Even China's shipbuilding industry, after nearly 2 decades of modernization and expansion, is hardly at a globally competitive level, technologically speaking. According to a Chinese study of its shipbuilding industry, while most advanced shipbuilding countries are at a level of integrated hull construction and outfitting, utilizing equipment- and information-intensive modular production techniques (Grade 4 capabilities) and are moving toward Grade 5 capabilities ("agile shipbuilding"), Chinese shipyards are largely stuck at Grade 2 (basic hull construction and pre-outfitting of metal ships), moving toward Grade 3 capabilities (limited modular construction involving hull block construction and zone outfitting).[73] Chinese yards still make only limited use of computers in ship design and for nesting, interference checking, and outfitting; consequently Chinese shipbuilders spend twice as long on design efforts as in the West. In general, a Chinese shipyard takes two to six times longer than advanced shipbuilding countries to construct a comparable ship. In addition, Chinese yards are deficient in several areas of ship construction, including mechanization and advanced welding technologies.[74]

In fact, the acquisition of advanced manufacturing technologies does not guarantee that they will be exploited to their full potential. For example, while the Chinese have invested heavily in acquiring sophisticated, numerically controlled and multiaxis machine tools, they are often underutilized. According to Pollack and Mulvenon, some large pieces of equipment—such as five-axis milling machines, stretchform presses, and autoclaves—are used so infrequently as to hardly justify their maintenance costs.[75] An eyewitness account of a visit to a Chinese armored vehicle factory mentioned an instance of workers milling engine parts by hand, even as several multi-axis machine tools lay fallow.[76]

At the same time, much of the Chinese defense industry continues to labor, with little success, to overcome its long-standing problems with quality control.[77] According to representatives of Western companies involved in joint ventures with the Chinese, quality assurance is still pretty much an "alien concept" to the Chinese.[78] Aviation factories are frequently described as dirty, poorly lit, heated and cooled, and generally disorganized.[79] Documentation and standardization—critical elements of quality assurance—are often lacking in aircraft production.[80] Within the shipbuilding industry, Chinese surface combatants sport "poor welding with signs of premature failure, inoperable equipment, and overall poor hull workmanship."[81] The *Luhu*-class destroyer contains limited damage-control facilities and basic design flaws in the weapons control room, exacerbated by poor construction.[82] Chinese-built frigates delivered to Thailand in the early 1990s featured hulls that buckled after firing trials of their on-board 5-inch gun, compartments without access, and doors and ladders that led nowhere;[83] in fact, upon arrival these ships had to be put into dry-dock for repairs and to improve their damage-control capabilities.[84]

It would also appear that commercial-to-military spin-on is considerably less than meets the eye. According to Wendy Frieman, despite years of official support of dual-use technology development, there is still little direct evidence of strong links between commercial S&T breakthroughs and the defense industrial base; in addition, it appears that the best people are **not** being diverted to military production, nor are current commercial technology acquisition efforts being executed so that the defense industry can benefit.[85] Most of the modern shipyards, aircraft plants, ground vehicle factories, and perhaps even missile facilities (i.e., those involved in building space-launch vehicles) appear to do little, if any, military production.

In addition, with few exceptions, spin-on is not a simple matter of "plug and play." In certain areas—particularly communications and information networking—the military can rather easily piggyback off of commercial off-the-shelf technologies (COTS). However, for most other applications—such as sensors, electronic warfare, and navigation/guidance—it can require a considerable systems integration effort to adapt a piece of commercial technology to military use.[86]

With regard to spin-on in the shipbuilding industry, Evan Medeiros has stated that "the conversion process has had little impact on improving China's ability to produce modern warships equipped with advanced naval technologies."[87] He adds that the opportunities for **direct** spin-on of civilian design and construction are limited: the shipbuilding industry's low technology base, while sufficient for cargo ship construction, offers little value-added to warship production. In particular, advanced naval designs require technologies and expertise, such as damage control and survivability, not accessible through civilian products.[88] These limitations, Medeiros argues, "will persist for years."[89]

Just as important, while defense conversion may lead to the acquisition of technologies that could be redirected toward military uses, the conversion process often acts as a distraction from defense production. China's shipyards have become so successful at commercial production that warship construction (measured in gross tonnage) is no more than perhaps 5 percent of all shipbuilding production sector output. As a result, low-profit naval construction is likely finding it increasingly difficult to compete with the more lucrative, export-oriented commercial shipbuilding sector.[90] It is likely that aircraft factories that have converted to civil aviation or ground ordnance plants that now build automobiles and motorcycles—especially foreign joint ventures that earn these enterprises hard currency—may also oppose efforts to divert assets to military production.

Finally, there is no assurance that defense conversion will even be a successful medium for spinning-on advanced technologies into China's commercial industrial base, let alone the country's military-industrial complex. In a study of 1,000 cases of civilian technology transfer to China, for example, only 200 came to any fruition; and of these, only half (100 cases) were deemed successful.[91]

Impediments to Technology Absorption and Exploitation.

All in all, China has experienced considerable and enduring problems "translating theory and design into reliable weapon systems."[92] Several factors contribute to the problems that the Chinese defense industry faces when it comes to expanding its ability to absorb and leverage new advanced technologies for armaments production. These barriers can essentially be broken down into three groups: technical, structural, and cultural.

Technical Deficiencies. China's defense industrial base lacks many of the basic technical skills necessary to fully exploit acquired technologies. One of the most critical of these is the defense industry's apparently enduring **weak systems integration capabilities**—the ability to envision, design, and develop a finished weapon system out of hundreds or even thousands of disparate components and subsystems and get it to function to its fullest potential as a single unit. According to a U.S. aviation industry representative, the Chinese "are especially deficient in systems integration" when it comes to aircraft manufacturing; he adds that "Chinese engineers and technicians are normally grounded in the basic discipline [of aircraft production], however, practical applications, manufacturing technologies, and overall experience are in short supply."[93] Pollack and Mulvenon assert that when it comes to the systems integration and engineering process, "the Chinese do not have a master plan that builds aircraft from the bottom up. Instead, they try to take parts off the shelf that were never designed to be part of any particular end product and try to make them fit."[94] Even the Cox Report concedes that China "lacks the ability to integrate the contributions of many disciplines that are required to utilize the rapidly emerging new technologies. The PRC system is unable to keep up with these basically new approaches."[95]

Ironically, some of this deficiency in systems integration/systems engineering on the part of the Chinese arms industry appears to be related to its "excessive focus" on technology acquisition over technology absorption. According to Mark Stokes, Chinese defense enterprises have traditionally purchased foreign equipment "without much thought as to how to integrate various components" into a single, workable system.[96] As such, arms producers may acquire the "know-how" but not necessarily the "know-why" of advanced weapons development and manufacturing.[97] As one European aviation industry representative put it, "Technology transfer is not an event, it is a process."[98]

Closely linked to the problem of poor systems integration and systems engineering skills is **limited workforce expertise** on the part of industry. According to one Western analyst, the Chinese traditionally have paid insufficient attention to training and workforce development.[99] Hence, advanced machine tools often go underused (or even unused), due to a lack of skilled operators.[100] In addition, factories often have so little actual contract work that many skilled workers gain only very limited experience with advanced manufacturing techniques; at the same time, many young bright and enthusiastic engineers and technicians have limited opportunities to apply their knowledge to actual programs.[101] For their part, state-owned enterprises (SOE) executives typically lack the managerial and entrepreneurial skills and experiences necessary to make market-oriented investment and production decisions.[102]

Structural Impediments. Another factor impeding the Chinese defense industry's ability to absorb and exploit new technologies is **overcapacity**. Quite simply, the country possesses too many workers, too many factories, and too much productive capacity for what few weapons it produces, resulting in redundancy and duplication of effort, inefficient production, and wasted resources.[103] For example, while Chinese aircraft production is estimated to require only about 200,000 workers, the aviation industry possesses a workforce nearly three times as large.[104] Within the shipbuilding industry, output is only 17 tons per person per year; in a comparable shipyard in a more advanced country, this figure is **700** tons per person per year.[105] And while this sector has enjoyed considerable commercial and financial success, it is beginning to fall victim to the consequences of rapid and uncoordinated overexpansion: as new ship orders have dropped and production has stagnated, yards are underbidding each other in an effort to win new contracts.[106] Nevertheless, China continues to add new dry-docks, heavy-lift cranes, and even a whole new shipbuilding facility (the ultra-modern Waigaoqiao yard, near Shanghai, due to open by 2000), which will nearly double the country's current shipbuilding capacity.

This overcapacity is exacerbated by an endemic **lack of sufficient capital**. The government often does not have enough money to put prototype projects into serial production.[107] In addition, most defense enterprises are starved of funds to underwrite plant modernization and to retain skilled workers; a newly hired university graduate entering the aviation industry, for example, earns less than 500 yuan a month—much less than he or she can make working in the coastal cities or for a foreign-run joint venture.[108] Low salaries are partly the reason for the exodus of skilled workers from China's new semi-conductor plant in Huahong.[109] Moreover, at least 70 percent of China's state-owned enterprises are operating at a loss—and the arms industries reportedly are among biggest money-losers.[110] As a result, most SOEs are burdened with considerable debt, much of which is triangular—that is, money owed to them by other unprofitable SOEs and therefore nearly uncollectable.[111] At the same time, most SOEs are still saddled with their *danwei* (work unit) obligations to pay for their workers' housing, health care, childcare, and retirement.[112]

The **highly compartmentalized, vertically integrated, and secretive nature of the Chinese arms industry** makes it difficult for arms producers to diffuse advanced technologies, share learning experiences, and collaborate on advanced weapons projects.[113] According to Pollack and Mulvenon, Chinese aviation enterprises typically find it difficult to communicate with each other horizontally, hampering "effective coordination;" as a result, firms will often not share information, even on joint programs.[114] This "stovepiping" has exacerbated the excess competitiveness and redundancy in manufacturing capability, since each defense enterprise often tries to "do it all," resulting in its acquisition of expensive machine tools that it then hardly uses (instead of subcontracting the work out to another firm).[115] Compartmentalization, poor interfirm linkages, and the lack of financial compensation particularly affects dialogue and exchange between the R&D institutes which design the weapons and the arms factories that produce these systems and keep all the profits. This process often results, according to one Chinese observer, in poor morale and motivation, the loss of talented workers, program delays, and poor quality assurance on the part of the design institutes, who "bear the heavy burden of continually breaking new ground but have nothing to do with the subsequent harvest."[116] Lines of communication between consumers of armaments (the PLA) and suppliers (the defense industry) are equally poor.[117]

Cultural Impediments. Finally, the **heavily centralized, hierarchical, bureaucratic, and risk-averse corporate culture** typically found within the defense industry makes it difficult to extract greater benefits from technology acquisitions.[118] In their recent study on Chinese capacities for innovation, Yuko Arayama and Panos Mourdoukoutas assert that "Chinese managers do not have either the will, the expertise, or the freedom to take the risks and make the adjustment associated with innovations."[119] Production management is often highly centralized and "personality-centric," with most critical project decisions being made by a single chief engineer. At the same time, lower-level managers "tend to be conformist, adhering to standard rules and procedures rather than to personal judgments based on their professional experiences."[120] Hence, they are usually reluctant to act on their own to deal with problems that might arise on the factory floor, inhibiting innovation and experimentation.[121] For example, Chinese technicians in the aviation industry lack practical experience in handling modern technology because of their "hesitancy of making a mistake and failing."[122]

The "SOE mindset" within the defense industry also undermines competitiveness and rationalization of arms production. For example, senior executives in China's military-industrial complex will often award contracts in order to preserve jobs and keep factories open, rather than on the basis of merit and competency.[123] For its part, the central government has traditionally been forgiving of bad performance within the SOE sector.[124]

Larry Wortzel, citing a U.S. aerospace industry representative, perhaps best sums up the impediments to innovation and technology exploitation arising from the defense industry's corporate culture:

> Part of the problem with Chinese [aircraft] manufacturing ... is that industrial management in China still relies on 1950s Soviet styles. This involves "batch-building" a full order of aircraft in advance based on state-planned and dictated order for parts and materials. As a consequence of this system, there are no direct lines of accountability for quality control, and no cost-cutting discussions or steps available to mid-level management. There is no competitive bidding for contracts, workers are redundant, and schedules continually slip because

state planning doesn't have a fixed required-delivery date for products . . . Young managers stay risk-averse and are reluctant to change or improve the system.[125]

Reforming the Defense Industry: Reorganization, Restructuring, and Rationalization.

China's current system, according to Arayama and Mourdoukoutas, "has yet to develop the entrepreneurial, managerial, and governmental regimes needed to master foreign technology and turn it into innovations that lead to sustainable competitive advantages."[126] The words, although aimed at the country's technological and industrial base in general, ring particularly true for China's defense industry. To be fair, the Chinese are certainly aware of the deficiencies in their defense industry and are undertaking efforts to improve their defense R&D and production processes. The current reform initiatives have their origins in the Fifteenth Communist Party Congress held in September 1997 and in legislation approved by the Ninth National People's Congress (NPC) in March 1998. The 1997 Party Congress laid out a radical agenda for restructuring and downsizing the SOE sector (including, of course, the defense industries) and for opening up SOEs to free-market forces—i.e., supply-and-demand dynamics, competition, and fiscal responsibility. The 1998 NPC refined this agenda by announcing plans to reorganize the government's defense industry oversight and control apparatus and establish new defense enterprise groups. At the same time, the NPC established a 3-year timetable for full implementation of these reforms.[127]

Reorganization. One of the most important actions to come out of the 1998 NPC was the dissolution of the old COSTIND. COSTIND was created in 1982 to centrally manage the process of research, development, testing, and production in the Chinese defense industry. In addition, it was intended to act as a bridge between the PLA and the civilian defense industries in order to ensure that the needs of the "consumer" were being heard and heeded; as such,

COSTIND was partly staffed by the military and headed by a PLA general. The self-evident failure of this organization, however, led to the spin-off of the military half of COSTIND into a new PLA General Armaments Department (GAD) and the creation of a "new," civilian-only COSTIND. The GAD will act as the purchasing agent for the PLA, overseeing defense procurement and watchdogging new weapons programs. Meanwhile, the new COSTIND will oversee defense-related S&T research, direct military R&D and production, and formulate and implement defense industry reform plans.[128]

Restructuring. Another key element of current defense reforms is the creation of ten new defense industry enterprise groups (see Table 1).[129] These enterprise groups replace the five old administerial-level pseudo-corporations—such as the Aviation Industries of China (AVIC), the China Aerospace Corporation (CASC), and the China North Industries Corporation (NORINCO)—that were essentially ministerial-level bodies directly answerable to the State Council. These new defense enterprises will be established as "genuine conglomerates," integrating research, production, and marketing.[130] At the same time, the government role in the daily operations of the defense industry will be greatly reduced,[131] and these new enterprise groups will have the authority to manage their own operations as well as the responsibility for their own profits and losses.[132] Current reforms include a significant downsizing of the defense industrial base, including worker layoffs and factory mergers and closures.[133] Both AVIC and the China State Shipbuilding Corporation (CSSC), for example, have announced plans to eliminate roughly one-third of their workforce by late 2000; NORINCO and the China National Nuclear Corporation (CNNC) have detailed lesser (although probably preliminary) layoffs in their industries (see Table 2).[134] At the same time, rationalization of the defense industry will likely include some factory closures as a result of government-encouraged mergers, as part of the policy—

enunciated at the 15th Party Congress in September 1997—of "letting the strong annex the weak."[135]

Old Corporate Entity	New Enterprise Group	Major Products
Aviation Industries of China (AVIC)	China Aviation Industry Corp. I	Fighter aircraft, bombers, transports, advanced trainers, commercial airliners
	China Aviation Industry Corp. II	Helicopters, attack aircraft, light trainers, UAVs
China Aerospace Corp. (CASC)	China Aerospace Science & Technology Corp.	Space launch vehicles, satellites, missiles
	China Aerospace Machinery & Electronics Corp.	Missiles, electronics, other equipment
China Ordnance Industry Corp. (COIC)/NORINCO	China North Industries Group Corp.	Tanks, armored vehicles, artillery, ordnance
	China South Industries Group Corp.	Miscellaneous ordnance, automobiles, motorcycles
China State Shipbuilding Corp. (CSSC)	China State Shipbuilding Corp. (Southern shipyards, based in Shanghai)	Frigates, smaller surface combatants, commercial ships
	China State Shipbuilding Industry Corp. (Northern shipyards, based in Dalian)	Destroyers, commercial ships
China National Nuclear Corp (CNNC)	China National Nuclear Corp.	Nuclear energy development, nuclear fuel and equipment
	China Nuclear Engineering & Construction Group Corp.	Construction of nuclear power plants, other heavy construction

Sources: *China Daily*, January 31, 1999; *China Daily*, February 5, 1999; Dow Jones International News Service, March 23, 1999; *China Daily Business Weekly Supplement*, May 9, 1999; *China Space News*, May 26, 1999; *AFP*, May 27, 1999; *Xinhua*, May 28, 1999; *China Daily*, July 2, 1999; *China Aero Information*, July 9, 1999; NORINCO website (www.norincogroup.com.cn); private conversations with Harlan Jencks and Mark Stokes.

Table 1. China Defense Industry Restructuring, 1999.

Defense Industry Reforms: Old Wine in New Bottles, or Rearranging the Deck Chairs on the Titanic?

So far, however, these reforms have been largely disappointing. For one thing, if the intention of creating these new industrial enterprise groups was to inject greater

Current Corporate Entity	Estimated Workforce, 1997	Projected Layoffs, 1998-2000
AVIC	560,000	150,000-200,000
CASC	270,000	None Announced
NORINCO	800,000	90,000*
CSSC	270,000	60,000-120,000
CNNC	280,000	30,000*

* announced so far; further layoffs anticipated

Sources: Harlan Jencks, *COSTIND is Dead, Long Live COSTIND!*; CNNC brochure; Airshow China '96 website; *China Daily*, October 3, 1997; *China Daily Business Weekly Supplement*, December 7, 1997; *Sing Tao Jih Pao* (Hong Kong), December 23, 1997; *Xinhua*, January 7, 1998; *Journal of Commerce*, December 19, 1998.

Table 2. Projected Layoffs in the Chinese Defense Industry.

competition into China's military-industrial complex—and therefore spur innovation and greater responsiveness to customer, i.e., PLA, demands—then these restructuring efforts have largely been a failure. At one time, it was expected that the Chinese would create large trans-sectoral, cross-competing defense conglomerates, similar to the South Korean chaebols or, more specifically, to horizontally integrated defense companies like Lockheed Martin or British Aerospace.[136] Such a strategy would have entailed a much more complicated restructuring of the defense industry, crafting enterprise groups that would have competed with each other to produce a broad array of weaponry. Instead, all Beijing did was break up each of its former defense corporations into two new groups.[137]

Even then, with few exceptions these new enterprise groups do not compete with each other directly. Of the two new corporations replacing AVIC, for example, all fighter aircraft and commercial jetliner production will be concentrated within one enterprise group, while all helicopter production will be centered in the other.[138] The nuclear industry will be split into separate enterprises for either construction or nuclear energy development.[139] NORINCO appears to be subdivided into one enterprise group mostly concerned with building tanks, armored vehicles, and other ground ordnance, while the other has been almost entirely civilianized, specializing in automobile and motorcycle production.[140] Even naval shipbuilding will see little direct competition between the two new corporations replacing the old CSSC: the new northern group will build destroyers, while the southern group will concentrate on frigates.[141] In fact, the Chinese appear to have intended that the new defense industries not vie directly with each other. For example, the two enterprise groups arising out of the CASC breakup "will not compete in terms of products," but rather, "in terms of their systems of organization and their operational mechanisms."[142] In addition, the new aviation groups were purposely "divided into conglomerates with different responsibilities."[143]

Rationalization of the defense industry has also been much slower than expected. AVIC, for example, was expected to lay off 60,000 workers in 1998; in reality, only 34,000 workers were let go.[144] At the same time, many of these layoffs have been illusory—often, displaced workers are transferred to "alternative" employment within the same corporation or other SOEs are forced to accept them, or they are simply paid to stay home.[145] At the same time, there have been no public announcements of any defense factories being closed or merged. In fact, it is increasingly likely—particularly as the Chinese economy continues to perform poorly—that the 3-year timetable for full implementation of these reforms will slip by several years and that many of the more drastic aspects of SOE reform (i.e., workforce layoffs and enterprise bankruptcies) will be mitigated for the sake of social stability.[146]

It is unclear how independent these new defense enterprises will be of government control or how responsible they will ultimately be for their own financial well-being. According to one Chinese publication, the two new aviation enterprise groups remain under the "direct leadership of the central government," and will be "operated as state holding companies."[147] Beijing has made it clear from the beginning of the reform effort that arms production is a strategic industry too critical to national security to be privatized.[148] The central government will keep the new defense enterprises under much stricter supervision than other types of reformed SOEs; these same rules, however, work in favor of the arms industries: Beijing will be pressured to continue subsidizing them in order to preserve key arms programs.

Finally, and perhaps most important, the reform initiatives announced so far do not directly address many of the impediments affecting technology absorption and leveraging—that is, the defense industry's weak systems integration/systems engineering capabilities, the dearth of workforce expertise, compartmentalization and redundancy, and corporate culture. As a result, it is doubtful that these reforms **as they currently stand** will go very far in injecting market forces that would, in turn, drive innovation and technology exploitation. Indeed, despite current defense industry restructuring, Beijing has already called for additional initiatives to "introduce administrative and management reforms to meet the demands of the socialist market economy."[149]

Conclusions.

China's military-industrial complex arguably has made some impressive gains when it comes to producing highly capable weapon systems. Beijing has a proven track record in the area of reasonably sophisticated tactical missile systems, particularly with regard to ballistic and antiship cruise missiles, and, to a lesser extent, surface-to-air missiles. From this, one could infer that China's missile industry does not suffer as much from the kind of deficiencies—weak systems integration skills, overcapacity, corporate culture, etc.—that retard technology absorption and exploitation in other defense industrial sectors. (It is perhaps revealing that, of all the defense sectors, only the missile industry has not announced any plans to lay off workers.) The shipbuilding industry has demonstrated some success in leveraging commercial technologies, particularly "soft" technologies like Western management techniques.[150] The military is almost certainly benefiting from the acquisition of dual-use technologies in such areas as telecommunications and information-processing, including fiber optics, cellular and wireless communications, computers, and the Internet. And despite delays, the country is progressing in the development of a modern, fourth-generation fighter aircraft.

One also should not rule out the potential long-term impact of direct foreign assistance on Chinese capabilities to leverage advanced technologies. In particular, the presence and involvement of Russian and Israeli advisers and technicians in Chinese arms factories and on key weapons programs have considerable potential to aid the Chinese in overcoming long-standing deficiencies in such areas as design, engineering, and systems integration. In addition, the growing use of foreign military technologies, through licensing and other types of technology transfer, offers considerable potential for advancing Chinese arms production. According to Arayama and Mourdoukoutas, given China's poor capacities for innovation, "the **only** rational choice for Chinese managers is imitation . . . that is, mass production of products invented and innovated elsewhere around the globe."[151]

At the same time, this progress must be qualified. For one thing, much of success within the missile sector is directly due to the fact that it, along with nuclear weapons and satellites, has been a national priority since the late 1950s.[152] As a result, the missile industry has been the beneficiary of considerable high-level support—in terms of both money and manpower—held constant over several decades; for example, two recent directors of COSTIND—Ding Henggao and Cao Gangchuan (who currently heads the GAD)—are both graduates of Russian missile academies.[153] Given its limited resources, it will be difficult for Beijing give equal time and attention to many other weapons programs.

Secondly, Chinese success in missile development could be a result of the fact that it is technologically not as challenging as other types of weapon systems. Missile production is not exactly rocket science any more—it requires the mastery of a relative handful of technologies that, although not easy to master, are nonetheless mature and increasingly proliferated: aerostructures, propulsion, guidance, and warhead (witness the fact that at least a dozen other developing countries have also developed their own ballistic missile systems[154]). Missile systems do **not** require the more complicated technologies typically found in combat aircraft, armored vehicles, and warships, such as life-support systems, communications and information processing, self-defense and damage control, long-life propulsion, and exotic materials. As a result, missile systems may be less a pocket of excellence than an "island of adequacy" surrounded by a sea of mediocrity.[155]

With regard to other defense sectors, Chinese progress in technology leveraging will continue to be less than optimal, resulting in a haphazard, piecemeal, and drawn-out military modernization process. At present, China's defense industry is still ill-suited to taking full advantage of many military, dual-use, and production-related technologies being made available to it. At the same time, current efforts to improve the environment for technology leveraging are inadequate. In short, the defense industry is broken, and the Chinese are not fixing it. Barring dramatic reforms, the future of China's military-industrial complex will most likely resemble the present: a handful of promising technology developments, impeded by delays, small and fitful production runs, and a steep learning curve. As a result, while China may experience some modest absolute gains in military capabilities, it appears to be achieving little by way of relative gains—that is, closing gaps in its military-technical position vis-à-vis its likely regional competitors, who are likewise striving to add new high-tech capabilities to their defense postures.[156]

Admittedly, adequacy may be sufficient for China to extract an impressive degree of **asymmetric** military capability from its missile industry. Beijing can and probably will attempt to leverage militarily its strengths in missile technology and production to compensate for weaknesses elsewhere. Nevertheless, a reasonably proficient missile industry is hardly an indicator of an overall advancing defense industrial base. Missiles can only add so much to Chinese military capabilities, and in most other areas—especially aviation, surface ships, and ground equipment—the PLA is still not getting much bang for its buck. Even in the missile sector, there remains considerable uncertainty as to its ability to extract much military capability from current technical inputs—particularly when it comes to radar-guided air-to-air missiles, land-attack cruise missiles, and other air-to-surface precision-guided missiles (PGMs).

This is not to say that China does not or cannot constitute a significant military challenge to the region. Even if perhaps 90 percent of the PLA remains a "junkyard army," size still matters, and the sheer mass of brute force is something Beijing still has in abundance. A large, relatively backwards armed force, combined with a limited number of indigenous military-technological breakthroughs and, in particular, a few well-targeted arms imports, can still have a significant impact on the balance of power in East Asia. Nevertheless, if China succeeds in becoming a major conventional military power, it will more likely be in spite of its defense industry than because of it.

CHAPTER 2 - ENDNOTES

1. The author is grateful to several China security and defense industry experts who, in private discussions, e-mail, and other fora, have offered considerable insights and commentary on Chinese defense industrial modernization efforts, in particular, Dennis Blasko, Shuhfan Arthur Ding, John Frankenstein, Bates Gill, Harlan Jencks, Evan Medeiros, James Mulvenon, and Larry Wortzel.

2. In particular, see Mark A. Stokes, *China's Strategic Modernization: Implications for U.S. National Security*, Carlisle, PA: Strategic Studies Institute, U.S. Army War College, September 1999.

3. Scott McMahon and Dennis M. Gormley, *Controlling the Spread of Land-Attack Cruise Missiles*, Marina Del Rey, CA: American Institute for Strategic Cooperation, January 1995, pp. 22, 25, 53, 81; David Fulgham, "Cheap Cruise Missiles a Potent New Threat," *Aviation Week & Space Technology*, September 6, 1993, pp. 54-55; John J. Fialka, "Poor Man's Cruise," *Wall Street Journal*, August 26, 1993, p. 1; Vivek Raghuvanshi, "India Adds Stealth to Top Fighters," *Defense News*, October 23, 1995, pp. 1, 44; Shirley Kan, *China: Ballistic and Cruise Missiles* (CRS), p. 18; U.S. Department of Defense, *Report to Congress Pursuant to the FY 99 National Defense Authorization Act*; "Cruise Missile Makers Proliferate," *Flight International*, July 28, 1999, p. 21.

4. U.S. Department of Defense, *Report to Congress Pursuant to Section 1305 of the FY 97 National Defense Authorization Act*; Website: "Chinese Military Aviation" (*http://www.fortunecity.com/tattooine/vonnegut/172/gallery.com*); *Jane's Defense Weekly*, December 23, 1998; David Silverberg, "Emerging Nations Hunger for Precision Weapons," *Defense News*, February 8, 1993, pp. 9, 44.

5. "Beijing Develops New Radar-Absorbing Materials," *Jane's Defense Weekly*, February 24, 1999; Fulgham, "Cheap Cruise Missiles a Potent New Threat," p. 55.

6. Stokes, *China's Strategic Modernization*, pp. 14-19, 109-134; "Panel Says Spying Aided Conventional Weaponry," *Aviation Week & Space Technology*, May 31, 1999.

7. *Cox Report*, part 2 (Internet version).

8. Stokes, *China's Strategic Modernization*, pp. 42-51; Solomon M. Karmel, "The Maoist Drag on China's Military," *Orbis*, Summer 1998, pp. 378-379.

9. Richard Fisher, "Foreign Arms Acquisition and PLA Modernization: Appendix," in James R. Lilley and David Shambaugh, eds., *China's Military Faces the Future*, Armonk, NY: M.E. Sharpe, 1999.

10. Bates Gill and Taeho Kim, *China's Arms Acquisitions From Abroad: A Quest for "Superb and Secret Weapons,"* Oxford: Oxford University Press, 1995.

11. Fisher, "Foreign Arms Acquisition and PLA Modernization: Appendix," pp. 131-135; Stokes, *China's Strategic Modernization*, pp. 79-86; Duncan Lennox, "China's New Cruise Missile Program 'Racing Ahead,'" *Jane's Defense Weekly*, January 12, 2000, p. 12; "Contracts in the Aerospace and Military-Technical Sphere Between Russia and the PRC," *Yadernyy Kontrol*, Moscow, July-August 1998, translated and reprinted in *FBIS*, September 23, 1998; Fulgham, "Cheap Cruise Missiles a Potent New Threat," p. 55.

12. Fisher, "Foreign Arms Acquisition and PLA Modernization: Appendix," pp. 135-137.

13. *Ibid.*, p. 140.

14. *Ibid.*, pp. 140-143; "Contracts in the Aerospace and Military-Technical Sphere Between Russia and the PRC."

15. "Russia, China Military Cooperation Viewed," *ITAR-TASS*, Moscow, August 25, 1999, translated and reprinted in *FBIS*, August 25, 1999; Fisher, "Foreign Arms Acquisition and PLA Modernization: Appendix," pp. 144-146, 149-151; "China-Assembled Su-27s Make Their First Flights," *Jane's Defense Weekly*, February 24, 1999, p. 63; Jim Mann, "U.S. Says Israel Gave Combat Jet Plans to China," *Los Angeles Times*, January 8, 1995; David Fulghum, "New Chinese Fighter Nears Prototyping," *Aviation Week & Space Technology*, March 13, 1995; Gill and Kim, *China's Arms Acquisitions From Abroad*, pp. 84-85.

16. Fisher, "Foreign Arms Acquisition and PLA Modernization: Appendix," pp. 151-154; "Russian Radar First for China," *Jane's Defense Weekly*, August 4, 1999.

17. *Ibid.*, pp. 161-164.

18. "Russia Helps China Take New SSNs into Silent Era," *Jane's Defense Weekly*, August 13, 1997, p. 14; Fisher, "Foreign Arms Acquisition and PLA Modernization: Appendix," pp. 165-166; *Jane's Fighting Ships, 1999-2000*, Coulsdon, UK: Jane's Information Group, 1999, p. 116; "Contracts in the Aerospace and Military-Technical Sphere Between Russia and the PRC."

19. Website: "Chinese Military Aviation," *http://www.concentric.net/~Jetfight/Z-8_Z-9_Z-11.htm*.

20. "Chinese AAM Aspirations May Build on Alenia Aspide," *Flight International*, November 27, 1996, p. 22; "China Displays Export Air Defense Missile," *Aviation Week & Space Technology*, December 2, 1996, p. 61.

21. Fisher, "Foreign Arms Acquisition and PLA Modernization: Appendix," pp. 161-162.

22. Wendy Frieman, "The Understated Revolution in Chinese Science and Technology: Implications for the PLA in the 21st Century," in James R. Lilley and David Shambaugh, eds., *China's Military Faces the Future*, Armonk, NY: M.E. Sharpe, 1999, pp. 247-267; Jonathan Pollack and James Mulvenon, *Assembled in China: Sino-U.S. Collaboration and the Chinese Civilian Aviation Industry*, ms., Santa Monica, CA: RAND, August 1998, pp. 28-36.

23. U.S. General Accounting Office, *Export Controls: Sensitive Machine Tool Exports to China*, GAO/NSIAD-97-4, Washington, DC: U.S. Government Printing Office, November 1996.

24. Briefing by the consulting group DFI International, April 1999; news release by Bombardier Aerospace, November 20, 1998; "Joint Ventures Star in Beijing," *Aviation Week & Space Technology*, October 16, 1995, pp. 22-23.

25. *Cox Report*, Parts 2 and 13, internet version; Nigel Holloway, "Cruise Control," *Far Eastern Economic Review*, August 14, 1997, pp. 14-16.

26. Evan S. Medeiros, *Linking Defense Conversion and Military Modernization in China: A Case Study of China's Shipbuilding Industry*, ms., February 1998, pp. 2-12.

27. Stokes, *China's Strategic Modernization*, pp. 29-31.

28. Frieman, "Understated Revolution," p. 10; "NEC starts semiconductor production in Shanghai," *Reuters*, February 23, 1999; "NEC begins production at joint chip firm in Shanghai," *Associated Press*, February 23, 1999.

29. James Cox, "Siphoning U.S. Companies' Knowledge," *USA Today*, February 16, 1996; Kenneth Timmerman, "China Shops;" *American Spectator*, March 1995; "China's Gates Swing Open," *Business Week*, June 13, 1994.

30. Wendy Frieman, "Arms Procurement in China: Poorly Understood Processes and Unclear Results," in Eric Arnett, ed., *Military Capacity and The Risk of War: China, India, Pakistan and Iran*, Oxford: Oxford University Press, p. 82, emphasis added.

31. U.S. Department of Defense, *Report to Congress Pursuant to the FY 99 National Defense Authorization Act*.

32. "Chinese Missiles Basic to New Strategy," *Aviation Week & Space Technology*, March 8, 1999, p. 59-60; *Reuters*, February 9, 1999.

33. Wei-Chin Lee, "The Birth of a Salesman: China as an Arms Supplier," *Journal of Northeast Asian Studies*, Winter 1987-88, p. 39.

34. Bates Gill, "Chinese Military Hardware and Technology Acquisitions of Concern to Taiwan," in James Lilly and Chuck Downs, eds., *Crisis in the Taiwan Strait,* Washington, DC: NDU Press, 1997, pp. 118-119.

35. "China Displays New, Old? Fighter," *Aviation Week & Space Technology*, November 23, 1998, p. 22; Richard Fisher, "Heritage Report on China's 1998 Zhuhai Air Show," http://www.heritage.org/exclusive/zhuhai/part2.html; *Xiandai Bingqui*, January 3, 1999.

36. Frieman, "Arms Procurement in China," p. 82.

37. Lennox, "China's New Cruise Missile Program 'Racing Ahead.'"

38. Fisher, "Foreign Arms Acquisition and PLA Modernization: Appendix," p. 131.

39. U.S. Department of Defense, *Report to Congress Pursuant to the FY 99 National Defense Authorization Act*; "Cruise Missile Makers Proliferate."

40. "PL-7," and "R550 Magic," *Jane's Air-Launched Weapons*; *China Today: Aviation Industry*, compiled by the editorial board of the China Today Series, Beijing: Tri-Service Press, 1989, pp. 303-305.

41. "New Chinese Surface-to-Air QW-2 System to Enter Production Soon," *Jane's Defense Weekly*, September 23, 1998.

42. "China Develops FM-90 SAM," *Jane's Defense Weekly*, December 23, 1998, p. 8.

43. Website: "Chinese Navy," http://www.fortunecity.com/tattooine/vonnegut/172/han_xia_kilo_song.htm; *Jane's Fighting Ships, 1999-2000*, p. 117; Antony Preston, "The Submarine Threat to Asian Navies," *Asian Defense Journal*, October 1995, p. 18; Robert Sae-Liu, "Submarine Force Priority for China's Modernization Plan," *Jane's Defense Weekly*, May 13, 1995, p. 18.

44. Huang Qiang, "Will China's Aviation Industry Be Able to Get Out of the Doldrums Soon?" *Keji Ribao*, Beijing, July 8, 1999, p. 8, translated and reprinted in *FBIS*, August 12, 1999.

45. Ken Allen, "PLAAF Modernization: An Assessment," in James Lilly and Chuck Downs, eds., *Crisis in the Taiwan Strait,* Washington, DC: NDU Press, 1997, pp. 240, 243; "China Displays New, Old? Fighter," p. 22; "Engineer Arrested for J-10 Leak," *Jane's Defense Weekly*, August 4, 1999.

46. "First Flight for F-10 Paves Way for Production," *Jane's Defense Weekly*, May 27, 1998, p. 17; "China Start to Flight Test New F-10 Fighter," *Flight International*, May 20, 1998, p. 5.

47. "Russian Imports Step in to Fill the Arms Gap," *Jane's Defense Weekly*, December 10, 1997, p. 28; Allen, "PLAAF Modernization: An Assessment," p. 244.

48. Allen, "PLAAF Modernization: An Assessment," p. 244; see also John Frankenstein and Bates Gill, "Current and Future Challenges Facing Chinese Defense Industries," *China Quarterly*, June 1996, pp. 411-415.

49. Frankenstein and Gill, p. 414; Gill and Kim, p. 45, 93, 142-143.

50. *Jane's Fighting Ships, 1999-2000*, pp. 119-120, 124-125.

51. *Ibid.*, pp. 117-118; website: "Chinese Navy," http://www.fortunecity.com./tattooine/vonnegut/172/ han_xia_kilo_song.htm.

52. "First Flight for F-10 Paves Way for Production."

53. Robert Sae-Liu, "Second Song Submarine Vital to China's Huge Program," *Jane's Defense Weekly*, August 18, 1999, p. 17.

54. "Aspide," *Jane's Air-Launched Weapons*; "9K111 Fagot Antitank Guided Missile System" and "HJ-8," *Jane's Infantry Weapons, 1999-2000*.

55. *Gong Shang Shi Bao, Commercial Times*, Taipei, August 24, 1999. I am grateful to Shuhfan Arthur Ding for bringing this article to my attention.

56. "First Flight for F-10 Paves Way for Production;" Fisher, "Heritage Report on China's 1998 Zhuhai Air Show," http://www.heritage.org/exclusive/zhuhai/part2.html, *Chuan-Chiu Fang-Wei Tsa-Chih*, Taipei, March 15, 1999; "Chinese Military Aviation," http://www.concentric.net/~Jetfight/J-10_J-11_FC-1.htm.

57. "China Considers Plans to Extend Fighter-Bomber's Capabilities," *International Defense Review*, January 1999, p. 63;

Douglas Barrie, "Chinese Turn to UK Firms to Fill Strike-Fighter Gap," *Defense News*, February 1, 1999, p. 6; Fisher, "Heritage Report on China's 1998 Zhuhai Air Show," *http://www.heritage.org/exclusive/zhuhai/part2.html*.

58. *Jane's Fighting Ships, 1999-2000*, pp. 117, 119-120.

59. Rupert Pengelley, "Grappling for Submarine Supremacy," *International Defense Review*, July 1996, p. 53; Robert Sae-Liu, "Submarine Force Priority for China's Modernization Plan," *Jane's Defense Weekly*, May 13, 1995, p. 18; Gill, "Chinese Military Hardware and Technology Acquisitions of Concern to Taiwan," p. 113.

60. "China Launches a Powerful New Super Warship," *Jane's Defense Weekly*, February 3, 1999, p. 16.

61. Larry M. Wortzel, *China's Military Potential*, Carlisle, PA: U.S. Army War College, October 1998, p. 18.

62. "Russia, China Conclude Deal to Produce Fighter Jets," *Interfax*, Moscow, in English, August 6, 1999.

63. During the 1970s, the Chinese reportedly acquired a few MiG-23s from Egypt, but a Chinese copy of the aircraft has never appeared in public. Richard A. Bitzinger, "Arms to Go: Chinese Arms Sales to the Third World," *International Security*, Fall 1992, p. 99.

64. Production of the Z-8 was delayed for several years during the early 1980s, probably because of engineering problems, and ultimately the Chinese never produced more than a handful of these helicopters. *China Today: Aviation Industry*, pp. 179-180. See also "Chinese Military Aviation," *http://www.concentric.net/~Jetfight/Z-8_Z-9_Z-11.htm*.

65. In the mid-1970s, China actually purchased a license to manufacture the Rolls-Royce Mk.202 *Spey* turbofan engine, which they designated the WS-9; in the end, however, even after 20 years of effort—and the import of hundreds of specialized tools and considerable technical assistance on the part of the British—only a handful of WS-9 engines were ever built. *China Today: Aviation Industry*, pp. 221-223; Wendy Frieman, "Foreign Technology and Chinese Modernization," in Charles D. Lovejoy, Jr., and Bruce W. Watson, eds., *China's Military Reforms: International and Domestic Implications*, Boulder: Westview Press, 1986, p. 59; "China Considers Plans to Extend Fighter-Bomber's Capabilities," p. 63; Barrie, "Chinese Turn to UK Firms to Fill Strike-Fighter Gap."

66. Larry Wortzel, remarks made at the RAND/CAPS conference on the Chinese military, July 1999.

67. Frieman, "Foreign Technology and Chinese Modernization," pp. 56-58; Norman Friedman, "Chinese Military Capacity: Industrial and Operational Weaknesses," in Eric Arnett, ed., *Military Capacity and The Risk of War: China, India, Pakistan and Iran*, Oxford: Oxford University Press, pp. 66-69; "China: Back to the Future," *Far Eastern Economic Review*, March 11, 1999, pp. 10-14.

68. Frankenstein and Gill, p. 414; see also Huang, "Will China's Aviation Industry be Able to Get Out of the Doldrums Soon?"

69. Quoted in Allen, "PLAAF Modernization: An Assessment," p. 241.

70. Pollack and Mulvenon, *Assembled in China,* pp. 9-12.

71. "China-Assembled Su-27s Make Their First Flights," *Jane's Defense Weekly*, February 24, 1999.

72. "PRC Plant to Start Su-27 Assembly in Two Years," *ITAR-TASS*, February 23, 1999; "China-Assembled Su-27s Make Their First Flights."

73. Rao Gangcan, *Development and Outlook in Newbuilding Technology in China*, ms., 1998, pp. 14-17.

74. *Ibid.*, p. 9-14.

75. Pollack and Mulvenon, *Assembled in China,* p. 28.

76. Wortzel, *China's Military Potential*, p. 19.

77. Bates Gill, "The Impact of Economic Reform Upon Chinese Defense Production," in C. Dennison Lane, Mark Weisenbloom, and Dimon Liu, eds., *Chinese Military Modernization*, London: Kegan Paul International, 1996, pp. 148-149.

78. Pollack and Mulvenon, *Assembled in China,* pp. 49-51; Wortzel, *China's Military Potential*, p. 18.

79. *Ibid.*, p. 29.

80. *Ibid.*, p. 40.

81. Medeiros, *Linking Defense Conversion and Military Modernization in China*, p. 13.

82. *Ibid.*, p. 14.

83. Friedman, "Chinese Military Capability: Industrial and Operational Weaknesses," p. 72.

84. *Jane's Fighting Ships, 1999-2000*, p. 698.

85. Frieman, "Understated Revolution," pp. 262-264.

86. Personal communication with a U.S. defense electronics expert, August 1999.

87. Medeiros, *Linking Defense Conversion and Military Modernization in China*, p. 20.

88. *Ibid.*, pp. 14-15, 19.

89. *Ibid.*, p. 20.

90. *Ibid.*, p. 19.

91. Dean Cheng, *Civil-Military Integration in the Chinese Aircraft Industry*, ms., April 1999, p. 6.

92. Stokes, *China's Strategic Modernization*, p. 136.

93. Quoted in Allen, "PLAAF Modernization: An Assessment," p. 241.

94. *Ibid.*, p. 236.

95. *Cox Report*, Part 13, Internet version.

96. Stokes, *China's Strategic Modernization*, p. 136.

97. Cheng, *Civil-Military Integration in the Chinese Aircraft Industry*, p. 6.

98. "Time Out in Asia," *Flight International*, November 5, 1997, p. 39.

99. Cheng, *Civil-Military Integration in the Chinese Aircraft Industry*, p. 6.

100. Allen, "PLAAF Modernization: An Assessment," p. 235.

101. Pollack and Mulvenon, *Assembled in China,* pp. 37, 47-48.

102. Yuko Arayama and Panos Mourdoukoutas, *China Against Herself: Innovation or Imitation in Global Business?*, Westport, CT: Quorum, 1999, pp. 69-82; "China: Back to the Future," p. 12; "The Crisis of State-owned Enterprises in Mainland China Worsens," *Cheng Ming*, Hong Kong, November 1, 1996, pp. 54-57.

103. "Time for a Reality Check in Asia," *Business Week*, December 2, 1996, pp. 60, 66.

104. *China Daily*, October 3, 1997.

105. Rao, *Development and Outlook in Newbuilding Technology in China*, p. 17.

106. Shen Bin, "Shipbuilding Giant Faces Division Plan," *China Daily Business Weekly Supplement*, May 9, 1999, pp. 3-5.

107. "China: Back to the Future," p. 11.

108. Huang, "Will China's Aviation Industry Be Able to Get Out of the Doldrums Soon?"

109. *Gong Shang Shi Bao, Commercial Times*, Taipei, August 24, 1999.

110. John Frankenstein, "China's Defense Industries: A New Course?" in James C. Mulvenon and Richard H. Yang, eds., *The People's Liberation Army in the Information Age*, Santa Monica, CA: RAND, 1999, pp. 197-199; Frankenstein and Gill, pp. 419-420; "Industry Embraces Market Forces," *Jane's Defense Weekly*, December 16, 1998, p. 28.

111. Harlan Jencks, "COSTIND is Dead, Long Live COSTIND! Restructuring China's Defense Scientific, Technical, and Industrial Sector," in James C. Mulvenon and Richard H. Yang, eds., *The People's Liberation Army in the Information Age*, Santa Monica, CA: RAND, 1999, p. 61; "The Crisis of State-Owned Enterprises in Mainland China Worsens," *Cheng Ming*, Hong Kong, November 1, 1996, pp. 54-57.

112. "China: Out of Business," *Far Eastern Economic Review*, February 18, 1999, p. 14.

113. Roger Cliff, "China's Potential for Developing Advanced Military Technology," in Anthony Lanyi and Kimberly Brickell, eds., *Institutional, Economic, and Organizational Basis of Military Capability*, College Park, MD: Center for Institutional Reform and the Informal Sector, University of Maryland at College Park, 1999, pp. 227-228.

114. Pollack and Mulvenon, *Assembled in China,* pp. 43-44.

115. *Ibid.*, pp. 22-23.

116. Huang, "Will China's Aviation Industry Be Able to Get Out of the Doldrums Soon?"

117. Cheng, *Civil-Military Integration in the Chinese Aircraft Industry*, p. 6.

118. Huang, "Will China's Aviation Industry be Able to Get Out of the Doldrums Soon?"; Jencks, "COSTIND is Dead," p. 6; Wortzel, *China's Military Potential,* pp. 19-20; Pollack and Mulvenon, *Assembled in China,* pp. 17-20, 39-42; Stokes, *China's Strategic Modernization*, pp. 135-136; Gill, "The Impact of Economic Reform Upon Chinese Defense Production," pp. 156-158.

119. Arayama and Mourdoukoutas, *China Against Herself*, p. 11.

120. *Ibid.*, p. 73.

121. Pollack and Mulvenon, *Assembled in China,* pp. 45-46.

122. Allen, "PLAAF Modernization: An Assessment," p. 242.

123. Pollack and Mulvenon, *Assembled in China,* p. 19.

124. "China: Out of Business," p. 15.

125. Wortzel, *China's Military Potential,* p. 20.

126. Arayama and Mourdoukoutas, *China Against Herself*, p. 11.

127. "State Enterprises Ordered to Reform by Millennium," *Reuters*, March 18, 1999; Jencks, "COSTIND is Dead," pp. 65-69; "Overhauling China Inc.?" *Business Week*, August 25, 1997.

128. Beijing Central People's Radio Network, March 30, 1998; Jencks, "COSTIND is Dead," pp. 65-66.

129. "Defense Industry Crucial," *China Daily*, July 2, 1999, p. 1. Some defense electronics production, which used to be conducted under the auspices the Ministry of Electronic Industries, has been subsumed into the new Ministry of Information Industries; consequently, much of the country's electronics production will likely remain directly managed by the central government; personal communication with James Mulvenon.

130. *Kuang Chiao Ching*, Hong Kong, April 16, 1998; *Jingji Guanli*, April 5, 1999; Beijing Central People's Radio Network, March 30, 1998; Jencks, "COSTIND is Dead," pp. 68-69.

131. Beijing Central People's Radio Network, March 30, 1998.

132. "Defense Industry Crucial," *China Daily*, July 2, 1999, p. 1; "Chinese to Restructure Shipyard Department," *Journal of Commerce*, December 19, 1998; Jencks, "COSTIND is Dead," p. 69.

133. Jencks, "COSTIND is Dead," p. 69.

134. Kuang Tung-Chou, "Zhu Rongji Says Joint Stock System is not Panacea for Rescuing State-Owned Enterprises," *Sing Tao Jih Pao*, Hong Kong, December 23, 1997, p. A2.

135. "Mergers Sharpen Firms for Market," *China Daily Business Weekly Supplement*, January 4, 1998, p. 1.

136. *Ibid*. In fact, such broadbased defense firms—producing everything from fighter aircraft to missile systems to defense electronics—are fast becoming the norm in the West, as firms expand their range of business activities, both to reduce their vulnerabilities to market downturns in any one defense sector and to take advantage of synergies in new weapons development.

137. "Five Major Military Industry Corporations to be Reorganized," *Wen Wei Po*, Hong Kong, May 4, 1999, translated and reprinted in *FBIS*, May 9, 1999.

138. *China Aero Information*, July 5, 1999.

139. "Nuclear Sector to Undergo 'Largest Ever' Restructure," *China Daily*, February 5, 1999.

140. NORINCO website, *www.norincogroup.com.cn*.

141. "Shipbuilding Corporation to Split Into Two Groups," *Xinhua*, May 28, 1999.

142. "Applying Technology to National Defense," *China Space News*, May 26, 1999, p. 1.

143. "China's Aviation Industry to be Split in Two Groups," *Dow Jones International News Service*, March 23, 1999.

144. Pollack and Mulvenon, *Assembled in China*, p. 16; *China Daily*, January 31, 1999. These less-than-anticipated layoffs in the defense industry mirror similar developments occurring throughout the rest of the state-owned enterprise sector. According to official sources in Beijing, China expects to eliminate only three million jobs in the SOE sector in 1999—less than half of the earlier target of seven million. "Shanghai Slump Forces Retailers to Shut Up Shop," *Financial Times*, July 12, 1999, p. 14.

145. Pollack and Mulvenon, *Assembled in China,* p. 16; Jencks, "COSTIND is Dead," p. 69.

146. "Jiang Zemin Presides Over Strategic Readjustment of State-Owned Enterprise Reforms," *Hsin Pao*, Hong Kong, August 20, 1999, p. 26, translated and reprinted in *FBIS*, August 22, 1999; "China: Out of Business," *Far Eastern Economic Review*, February 18, 1999, p. 11; "China: What's Going Wrong?" *Business Week*, February 22, 1999, p. 49-50.

147. *China Aero Information*, July 5, 1999.

148. "Government Not to Interfere in Enterprise Mergers," *Zhongguo Tongxun She*, Hong Kong, September 18, 1999, translated and reprinted in *FBIS*, September 19, 1999; "Jiang Zemin Presides Over Strategic Readjustment of State-Owned Enterprise Reforms"; *Hong Kong Standard*, September 15, 1997.

149. "Jiang Urges Military Scientists to Contribute to Defense," *Xinhua*, July 1, 1999.

150. Medeiros, *Linking Defense Conversion and Military Modernization in China,* p. 12.

151. Arayama and Mourdoukoutas, *China Against Herself*, p. 11. Emphasis added.

152. "CAS Reveals Past Role in China WMD Programs," *Kexue Shibao*, Beijing, May 6, 1999, translated and reprinted in *FBIS*, July 22, 1999.

153. I am grateful to John Frankenstein for pointing this out.

154. Argentina, Brazil, Egypt, India, Indonesia, Iran, Iraq, Israel, North Korea, South Korea, Pakistan, and South Africa, according to the Center for Nonproliferation Studies at the Monterey Institute of International Studies.

155. Wortzel, *China's Military Potential*, citing a U.S. aircraft corporation representative, p. 20. A similar observation was made by Dennis Blasko, personal communications with the author.

156. See, for example, Richard A. Bitzinger and Bates Gill, *Gearing Up for High-Tech Warfare? Chinese and Taiwanese Defense Modernization and Implications for Military Confrontation Across the Taiwan Strait, 1995-2005*, Washington, DC: Center for Strategic and Budgetary Assessments, 1996.

CHAPTER 3

PLA LOGISTICS AND DOCTRINE REFORM, 1999-2009

Lonnie Henley

The Chinese People's Liberation Army[1] (PLA) has implemented a number of reforms in the past 2 years intended to improve its ability to conduct mobile warfare. The main impetus is the realization that the PLA is not agile enough to cope with a fast-moving, modern opponent even on its own home turf. In order to address these weaknesses, the PLA is seeking to improve both its logistics support structure and its operational doctrine. Even if successful, these reforms may not significantly increase China's power projection capability, but they will improve its ability to move and sustain forces within China and around its periphery.

Two recent programs seem particularly important. One encompasses a group of decisions aimed at standardizing military operations and training. Chief among these was the issuance in January 1999 of new "combat regulations" or "operational ordinance" designed to standardize PLA doctrine, tactics, techniques, and procedures for combat operations.[2] The second is a 10-year effort to restructure the logistical system of the entire PLA announced by the General Logistics Department director, General Wang Ke, in November 1998. If it is successful, logistics reform will have a significant impact on China's military capabilities, as the inability to sustain large forces in intense, fast-moving combat operations is among the PLA's greatest weaknesses.

Logistics Reforms.

Wang Ke announced the logistics reform program at a "Forum on the Features and Rules of Logistics" in November 1998, and it was officially enacted at the expanded meeting of the Central Military Commission in December 1998. It was widely publicized in following months, including special expanded meetings of the General Logistics Department (GLD) in Beijing and logistics departments in the military regions and Service headquarters.[3]

The logistics program was also encompassed in the major reform document of the late 1990s, PLA Combat Order No. 13, "PLA Joint Campaign Program," January 24, 1999, which is discussed in more detail below. The GLD issued an All-Army Joint Logistics Implementation Plan shortly thereafter, and the organization and planning phase of the reforms got underway in the spring of 1999.[4] This phase was to continue into early 2000.

GLD Chief Wang Ke listed the following objectives:

- integrate logistics for the three services;
- supply work for units should be standardized;
- convert officers' perquisites into cash allowances;
- "socialize" logistics services [in other words, out-source support functions to civilian contractors and society at large]; and,
- make management more professional and "scientific."[5]

Although not included on this list, another central objective cited elsewhere is to improve mobile logistical support for units operating away from their home areas.

Trial Programs. As has been PLA practice in the 1990s, the GLD first set up large-scale trial programs in several different regions to develop the details and work out practical considerations before the final decision was made. Wang cited experimental programs in Shenyang, Jinan, and Hainan beginning in 1998, and also pointed to 1998's massive flood relief efforts as contributing greatly to the understanding of joint service logistical operations.[6] The experimental programs did not include reorganization or restructuring, but focused on single-Service facilities providing vehicle repair, communications equipment repair, medical service, and supply to other Services.

Experimentation has continued since the formal decision in December, as the regions work out how best to implement the new approach. Beijing Military Region (MR) instituted a pilot program contracting out support functions for four hospitals, one group army, and five division-level units.[7] Nanjing MR has begun relying on civilian supply sources for vehicle parts.[8]

Joint Service Logistics. At both the national and the local level, the separate Services of the PLA—Navy, Air Force, and Second Artillery (the strategic missile forces)—have maintained separate logistical infrastructures since they were created in the 1950s. (There is no separate ground component headquarters in the PLA. The four General Departments serve as both army staff and joint staff.) The military regions' logistics departments supported the Army ground forces, as well as the organizations of the "institutional" PLA—academies and schools, think tanks, military region, district, and sub-district headquarters, etc. The MRs exercised nominal control over the other Services' logistics structures in the region, but actual control centered in the Service headquarters in Beijing.

This system has become increasingly unsatisfactory, for both operational and economic reasons. Lieutenant General Jiang Jiesheng, commander of the GLD's Wuhan Rear Base, outlined the operational imperatives in a February article in *Jiefangjun Bao*:

> Joint operations have become a principal form, and war is manifested as confrontation between systems. The

operational system has to be very much complete and in harmony as a whole, and the logistics support system has to work as an entity. Hence, implementing a joint logistics support system among the three armed services has become a historical necessity.[9]

Wang Ke characterized the old system as a relic of the centrally planned economy, now woefully inefficient and out of step with the rapid transition to market-based structures in the past 2 decades:

> The present logistics and supply system in the army was established in the early 1950's, with the Army, Navy, and Air Force operating their own vertical supply systems under the leadership of the General Logistics Department. This matched China's highly centralized planned economic setup and the rather low level of army modernization at that time. As China's socialist market economy and the army's modernization have developed, the defects of this logistics and supply system have become more and more evident; for instance, for the Army, Navy, and Air Force to independently organize their own supply does not match the demand of combined operations using a variety of weapons in modern war; there is a serious degree of duplicate construction, and supply efficiency is low and fails to meet the demand of building the army's quality; and decentralization of powers with too many heads facing the market cannot produce efficiency in scale.[10]

Under provisions of the reform program, those support functions that are common to all Services are being merged into new Joint Logistics [Sub-] Departments (JLD) in each of the military regions' logistics departments. The services will retain separate logistics structures for supplies and support unique to their own systems. Under the rubric of "network-style zoned supply work," the services are surrendering hospitals, medical supply and maintenance units, fuel stocks and distribution networks, general supply warehouses, vehicle supply and maintenance units, and some general transportation units to the control of the MR JLD. The reorganization is to be completed by early in 2000.[11]

Role of the Military Regions. Some Western observers think the demands of modern, high technology warfare make the MR structure obsolete, and have speculated that they may be abolished in the near future. The logistics reform program, however, explicitly calls for strengthening rather than reducing the role of the MRs and military districts. With regard to joint logistics structures, Wang Ke said,

> We should build a logistical system based on the military districts [*jun qu*] with joint supply for the Army, Navy, and Air Force.... In instituting joint logistics for the three services, proceeding from the current reality of the armed forces, it is necessary to establish a new-style logistics system based on the military district....[12]

(*Jun qu*, "military district," is probably understood to include *da jun qu*, which we translate as "military region.") *Wide Angle* quoted General Wang, and added,

> This means that the country's military regions will remain structurally unchanged for a long time to come and the implementation of a new-type logistic support system capable of simultaneously catering to the needs of the three services will raise the position of the country's military regions.[13]

Xinhua's (New China News Agency) report on the reforms said,

> The ongoing reform of linking the logistics work of the three services is based on major military regions, which coordinate the reform and take charge of the logistics supply of the materials commonly used by the three services in the relevant theater, thereby changing the independent and decentralized support system in the past.... Some Navy and Air force hospitals as well as fuel and military supply warehouses were handed over to major military regions.[14]

"Socialized Logistics." In a major departure from the PLA's tradition of self-reliance, the reform calls for the military to turn to the civilian market economy as much as possible for support functions, in an approach dubbed "socialized logistics." ("Socialized" referring to society, not socialism.) In part, this is an economic move, seeking to capitalize on the efficiency of the market and reduce the overall cost of support functions. But the main thrust is to relieve military units of unnecessary overhead and allow them to focus on their operational responsibilities. This is a major departure from the Maoist tradition of minimizing the Army's burden on the civilian society, dating back to the Eight Points of Attention and the Yan'an Spirit of the anti-Japanese and Civil War era. Apparently anticipating resistance to this change, the leadership has devoted some effort to explaining and popularizing the new approach. From *Xinhua* in June 1999:

> In the past, the armed forces undertook too heavy a burden by running their own kindergartens, schools, shops, restaurants, laundry shops, bathhouses, canteens, heat supply centers, and cleaning business, taking much of their military and civilian personnel's time and effort.[15]

From *Wide Angle* in March 1999:

> After shaking off a heavy burden of logistic support, the country's military regions and the locally stationed ground, naval, and air units will be able to devote more resources to both military training and equipment management and this will definitely help enhance the army's capability of fighting and winning a hi-tech war.[16]

From *Jiefangjun Bao* in February 1999:

> [A]n open equipment and technical support system takes the military self-support system as the system border, and takes society's resources as the system environment. It allows extensive exchange of information, material, energy and capital between the system and the environment, making full use of society's material, human and technical resources, strengthening the armed forces equipment and technical support ability and economically, reliably and efficiently fulfilling the military's weapons, equipment, and technical service requirements. . . . An open equipment technical support system is the effective measure to "contain the army in the

people, unite in peace and war." Once indicators based on market information of various kinds of essential elements of support power latent in society are understood, then it is possible to reduce unnecessary storage and maintenance ability in the system, and decrease the expenditure of the system. An open system can raise equipment technical support "resilience," which is good for lowering peacetime expenditures and can satisfy great needs in time of emergency, adapting better to "both fields."[17]

Socialized logistics will contract out most of the day-to-day service functions that sustain the PLA in garrison. Besides the items listed by *Xinhua*, other articles have mentioned fuel supply, vehicle spare parts, garbage collection, tree planting, child care, and management of officers' housing. Outsourcing of support functions will start with garrison units and PLA institutes in major cities, and gradually expand to include the entire force.[18]

It appears that this shift does not include agriculture and light manufacturing, the so-called "sideline production" which occupies another large part of PLA manpower. There are frequent references to Jiang Zemin's 1990 statement that the PLA must "eat imperial grain" and not follow the path of feeding itself, but always in a context that is clearly a metaphorical reference to the divestiture of commercial activities supplementing the PLA budget, not to actual food supplies. Similarly, Wang Ke's December article says in part,

> With regard to the question of army engagement in production operations, comrade Deng Xiaoping pointed out back in 1978 that units below army level must absolutely not run factories, and this road must be cut off. Those run for profit must all be shut down, no matter what reason is given.

Again, however, Wang focuses on for-profit operations, and not on factories that make uniforms, boots, etc., for the PLA's own use.

CMC Vice Chairman General Zhang Wannian clarified the issue in a speech in April 1999. Although he made no reference to the socialized logistics program, which was well underway by then, he quoted explicit statements by Jiang Zemin and emphasized that PLA agriculture and light manufacturing are not affected by the ban on PLA business enterprises:

> Chairman Jiang pointed out: "The agricultural and sideline production of the armed forces is an old tradition and is the Nanniwan spirit, which we must carry forward, consolidate, and improve.... It is necessary for the armed forces to run some agricultural farms, pigsties, and chicken farms to improve the living standard of the soldiers. We have all along been doing this. In the future, we shall continue to do so and should do an even better job in this respect."[19]

One might argue that food is the most fungible of goods which the civilian market in China is well able to supply, and that agricultural production is one of the activities most disruptive of military readiness and training. Obviously, those are secondary considerations against the PLA's deep-rooted, almost sacred tradition of self-reliance.

Mobile Logistics. As the PLA has come to grips with the requirements of Modern Local War under High-Tech Conditions, it has become clear that the static, fixed-depot structure of its logistical system is one of its greatest weaknesses. In order to cope with a fast-moving, modern military opponent, the PLA needs to be able to rapidly shift its elite units across China to the site of the conflict, and sustain them in combat without making them pull back to the nearest depot for repair or resupply. Throughout the 1990s, the ground forces and the PLA Air Force (PLAAF) have designated rapid response or "fist" units, and they and the Second Artillery have practiced rapid deployment some distance away from their home base for field training exercises. Only a few units have done such deployments, and with the possible exception of Second Artillery units, seldom cross MR boundaries. But this is a major focus of PLA modernization, and the logistics reform program seeks to restructure tactical and operational level logistics elements to provide better support for out-of-area deployments. The goal is that "military provisions should arrive before the combat units arrive, and this is a necessary condition for unfolding a combat operation."[20] This is certainly a minimum requirement if the PLA is to develop any serious mobile warfare capability.

The PLA is not abandoning its fixed-site logistics structure, but it is seeking to graft on a mobile structure to provide forward support to the maneuver units. Components of the mobile logistics effort include repair at forward positions, without evacuating equipment to the depot or factory level; more responsive and more mobile resupply arrangements; new organizations to support out-of-MR air and ground units; and underway replenishment for naval forces. All these concepts have been themes of logistics work in the PLA for several years, and the 10-year reform program will build on previous efforts.

The MRs have experimented with a variety of organizational structures in recent years to provide mobile support, with Beijing MR reportedly in the vanguard.[21] The current focus is on two types of organizations: emergency support brigades, consisting of a medical battalion, a supply battalion, a fuel battalion, and a transportation battalion; and emergency military depots, encompassing ammunition depots, fuel depots, general supply warehouses, field clinics, and transportation units.[22] Experimentation with emergency support brigades began as early as 1994, when such a unit was formed by a logistics sub-department in Guangzhou MR.[23] There has been a spate of articles praising support units that are able to conduct rapid repairs on the battlefield or during convoy movements, without evacuating systems to depot level.[24] The PLA Navy (PLAN) has experimented with emergency support detachments (*fendui*) to repair warships at sea rather than sending them to factories for repair.[25] The PLAAF has experimented with composite mobile support detachments, assembled of elements from several different organizations, to accompany fighter units deploying to airfields along the southeast coast.[26]

Standardization of Supply Procedures. The farther a PLA unit deploys from its home base, the more difficulty it is likely to have interacting with the support infrastructure at its deployed location. Stock numbers, requisition procedures, inventory standards, and other details of supply operations vary from region to region, creating a major hindrance to the PLA's ability to deploy units across China. This may account in part for the rarity of cross-region mobility exercises. Although there are few details on how GLD intends to address the problem, Wang Ke cited standardization as a major priority of the reform program:

> [W]e must gradually establish a set of supply standard systems and corresponding rules and regulations that are complete, dovetail with each other, and are scientific, rational, simple, and easy to implement, and implement scientific management and adherence to law in army logistical work.[27]

Monetization of the Officer Compensation System. Another theme of logistics reform is the conversion of officer perquisites into cash allowances, reflecting broader trends in the economy as a whole. The daily life of ordinary Chinese has changed dramatically in the past 20 years. In Maoist China, both rural and urban residents received most of their daily needs from their work unit, directly in the form of locally produced products, or indirectly in the form of ration coupons to exchange at state-owned stores and distribution points. Housing, education, health care, rail tickets, cooking fuel, and most other goods and services were either provided in kind, or heavily subsidized by the state. The transition to a market economy is converting almost all of those exchanges to cash transactions.[28] Cash in the pocket, and a choice of things to spend it on, creates a great deal of personal freedom and encourages more efficient allocation of goods, but the change also places a major strain on employers including the PLA. Military personnel costs have skyrocketed, as the PLA has to provide its personnel, especially officers, with ever-increasing volumes of cash to pay for goods and services. Chinese officers assert, though detailed figures are hard to obtain, that the PLA's budget increases over the past 10 years have been largely consumed by this one requirement.[29] The monetization process will continue under the logistics reform program, with particular emphasis on housing and vehicles. Construction and maintenance of officer housing will meanwhile be transferred to civilian government programs, as part of the "socialization" of logistics.[30] If successful, officers will buy or rent their housing on the private real estate market. Success depends to a great extent on the broader effort to privatize urban housing throughout China, however, and that is a huge program with uncertain prospects.[31]

Housing is not the only part of the compensation system that the PLA is trying to off-load onto civilian government. Wang Ke cites an August 1998 "insurance decision for military personnel," under which responsibility for medical and pension costs for veterans killed or injured in the line of duty is being shifted to the state insurance system—in other words, to local governments. The March 1999 article in *Wide Angle* discusses the intent to shift most of the "social insurance" and "social support" burden for military personnel to local authorities. The article confidently predicts that the PLA can "mobilize the enthusiasm of the local governments" for accepting this burden, because the divestiture of PLA commercial operations will leave those now-private enterprises highly "transparent," and allow local government enterprises to compete more effectively against them. One needn't be a hardened skeptic to question whether localities will be quite so enthusiastic about this shift. The author also characterizes the change as an issue of equity within the PLA; the benefits derived from PLA enterprises went to the small number of officers involved in those activities, while this reform will benefit "the broad masses of officers and men."

The primary advantage of providing officer benefits in cash rather than in kind, besides shifting some of it out of the military budget, is that the PLA can shed some of the manpower and overhead associated with housing construction and maintenance, medical care, and personal services such as drivers and household servants.

Not directly related to the monetization of officer perquisites, but in pursuit of similar objectives, is the establishment of new ranks and pay scales for non-commissioned officers in the July 1999 revision of the Soldiers' Service Regulation. The regulation creates six new grades, NCO 1 through 6, with a pay scale that increases in line with rank and term of service.[32]

Other Logistics Changes. There have been several other aspects of logistics reform mentioned in the press over recent months.

• Funding for logistics. One article cites the divestiture of PLA commercial enterprises as being directly tied to increased funding for maintenance within the PLA:

> In the past, some army units carried out commercial activities on the pretext of raising maintenance funds for the grassroots military units. However, only a small part of the funds raised by the units concerned had actually gone to the grassroots units.... [the CMC] came up with a decision in the second half of 1998 on banning the army and the armed police from commercial activities within a short period of time and increasing the army's maintenance funds in a timely fashion and adopted resolute measures to this end....[33]

• Skip-echelon logistics. Several articles mention streamlining logistics operations within the MR, allowing direct links between the MR Joint Logistics Department and regional level support units, on the one hand, and division, brigade, or regiment level units on the other.[34]

• Food service reforms. These include increasing the standard dietary allowance for soldiers, with more meat and greater variety (but no increase in per-soldier budgetary allowance for food); improved field kitchens, to better support mobile operations; issuing combat rations of toasted rice and hard tack instead of uncooked rice, to improve survivability on the battlefield by reducing the need to light fires in order to eat; and developing frozen and packaged foods for the destroyer *Qingdao* and training ship *Shichang* for their foreign port visits last year, with increased "freshness" so that sailors would not have to subsist for weeks on the standard PLAN at-sea rations of instant noodles and canned food.[35]

• Emphasis on faster throughput by the transportation system. Again, the PLA press cites the 1998 flood relief operations as having provided valuable real-world experience in rapid resupply operations. An interesting application of these efforts is a focus on faster intermodal transfer of materiel (from land and air, to sea modes) in support of joint service operations offshore and along the coast near Xiamen.[36]

• Improvement of logistics structures in the reserve forces. Fujian MD recently created a reserve force logistics brigade. Hubei MD is restructuring its reserve bridging units, in part to provide better support in future civil flood relief operations, while arguing with local governments over who should pay for such operations. Xinjiang MD is increasing its investment in dual-use transportation infrastructure, especially roads and service stations along the Tarim Desert Highway.[37]

The logistics reform program seems to represent a wide-ranging change in the structure and philosophy of PLA logistical support, based on 5 years of planning and experimentation. The official program has just begun, and it will be several years before we can assess whether it is achieving much progress.

Standardization.

Most of the other reforms instituted over the past year or two can be understood as an effort to create more standardization within the PLA. Obviously, no large military can be completely standardized. Units will always be tailored to their mission and local terrain, or differ just because that's how they developed over the years; the equipment inventory will always vary from unit to unit, because no large military can modernize all its equipment at once, and older systems retain some utility for some missions; commanders and staff will always interpret training standards and operational doctrine in their own way.

Within these inevitable limits, however, those units that deploy and fight away from their home region need to share key operational approaches and concepts with the units fighting alongside them. Standardization of supply and support procedures is important in this regard, as discussed above. Equally important is some uniformity in how officers are trained, and the development of the professional military education system over the past 20 years has contributed significantly to this goal.[38] In a military where most officers spend their entire career in one division or even one regiment, it is useful to bring them together on a regional or national level a few times in their career to study concepts and techniques of operation. But the most important consideration is that units from different parts of China must share the same understanding of how to fight and how to operate on the battlefield if they are to merge seamlessly when brought together for combat.

Operational Doctrine. In January 1999, CMC Chairman Jiang Zemin signed an order implementing the first major revision to PLA operational doctrine since the mid-1980s.[39] The old regulations, now referred to as the "first-generation" effort to standardize PLA operational procedures, did not address joint service operations or warfare under high-technology conditions, did not cover the full range from national- and campaign-level down to unit-level operations, and obviously did not incorporate the lessons of the Gulf War and other 1990s campaigns or evolving concepts referred to as the "revolution in military affairs."[40] The 13 new combat regulations, known collectively as the "Chinese People's Liberation Army Joint Campaign Program," are designed to rectify these omissions.

The new regulations, also called "operational ordinances," are the embodiment of 7 years of experimentation and reform in the PLA. From 1992 to 1995, the PLA conducted a series of what Americans call warfighting experiments. A number of units throughout the PLA were assigned to work on different aspects of modern warfare—night fighting, electronic warfare, battlefield mobility, strategic mobility, counter-reconnaissance and counter-surveillance, etc. Since 1995, we have seen several large-scale field-training exercises and numerous smaller exercises that attempted to pull together these separate components into coherent "methods of operations." The development of these methods has been a constant theme in the Chinese military press in the late 1990s.

Now we are told that during this same period, 1995 to late 1998, a body called the PLA Combat Regulations Compilation Committee has been coordinating a PLA-wide effort to codify the doctrinal concepts developed in these two phases of field experimentation. This committee included staff officers from all Services and all echelons, as well as researchers from PLA academies and think tanks. After extensive coordination and four meetings with the General Departments, the committee submitted the regulations to the CMC for approval late last year.[41]

There has been no explicit statement yet of the content or even titles of the 13 regulations, but one report says they are grouped into two categories: campaign regulations, further grouped into joint service campaigns and single-Service campaigns; and combat regulations, including "general regulations, the regulations for the combat operations conducted jointly by the units of different services, and the regulations for the combat operations conducted jointly by the units of different arms and different specialties." (The latter seems to refer to what the U.S. Army calls combined arms operations—infantry, tank, artillery, etc., working together—rather than joint-service operations.) The report later lists areas the regulations focus on, which may correspond to the 13 separate regulations. These areas are campaign principles; combat principles; campaign modes; combat modes; basic tactics; command system; logistics support; equipment support; political work; information warfare; electronic warfare; air defense; and countermeasures against the enemy's high-tech arms.

More information and time are required before we can judge the impact of these new regulations on PLA operational capabilities. At a minimum the regulations embody nearly a decade's worth of PLA thought on how to implement the general concept of "modern local war under high-tech conditions." They will certainly focus on the same concepts we have seen in PLA exercises over the last 4 years: progress toward joint operations; the expectation of a high-tech enemy, a focus on innovative tactics to neutralize the opponent's advantage; and efforts to create a more realistic training environment.

It is also likely, however, that except for those forces tagged for a potential conflict with Taiwan, the regulations will display the same muddle concerning strategic objectives that have characterized PLA modernization efforts in recent years. China has no major ground threat since the end of the Soviet Union and the collapse of Russia's economy, yet it retains one of the world's largest armor and mechanized force structures. Much of the PLA's experimentation since 1992 has focused on a large ground fight against a sophisticated opponent. Yet the likelihood of the U.S. Army invading China, or of anyone else being able to even contemplate such an attack, must seem vanishingly remote even to the PLA. China already possesses an enormous military advantage over its land neighbors, at least for a conflict within reach of its MR-based logistical network. Training to confront a high-tech opponent does make China's ground forces better, but without a ground threat to the homeland, it does not increase China's useable military capability, unless the PLA builds a logistics structure to project and sustain those forces outside China—and there is no evidence Beijing wants or intends to build such a structure.

Military Terminology. Standard doctrine requires a standard vocabulary. A new edition of *Chinese PLA Military Terminology*, China's equivalent of the U.S. *Dictionary of Military and Associated Terms (Joint Publication 1-02)*, was issued in 1998, replacing the previous 1982 edition. Most of the jargon associated with high-tech warfare has come into use in China since 1982, and the need for standardization is clear to anyone trying to make sense of exactly what a PLA publication is talking about. The Jinan MR newspaper cites the new dictionary, disseminated down to company level, as laying down standards for PLA units in the same manner as do the *Common Regulations*.[42]

Training Procedures and Evaluation Standards. Another effort at standardization focuses on the content of field and garrison training and on the criteria for evaluating such training. The PLA General Staff Department (GSD) Training Department has been issuing army-wide Master Training Plans for the past several years. New evaluation standards for both unit and individual training were promulgated by GSD at an army-wide training reform meeting in late 1997, under the title of "Provisional Stipulations for Grading Military Training." The directive applied to regimental and lower level unit training that year, and to division level training starting in 1998.[43] There has been no detailed reporting, however, on the content of the training program or on the evaluation standards.

Reorganization.

As of mid-1999, the PLA is engaged in a widespread restructuring of combat units and noncombat organizations and institutions. As the PLA leadership warned in early 1998, this phase of reorganization involves the "abolishment, merger, downgrading and reorganization of units and the departure of individuals."[44] Only the broad outlines of this reorganization are yet available, but some basic themes are apparent. One is a straight manpower reduction, with the focus on reducing the proportion of light infantry among the regular PLA forces. So far, some 14 light infantry divisions have been transferred to the People's Armed Police, and a number of other units have been reorganized to concentrate available major systems—tanks, infantry fighting vehicles, artillery, etc.—into smaller units. Some maneuver divisions are being reorganized as brigades, and some artillery regiments have been upgraded to brigades with the transfer of equipment from other, disbanded units.[45]

Another theme of structural reorganization is the streamlining of the "institutional" PLA: military region headquarters, garrison units, academies and schools, research institutions, etc. The reorganization started with issuance of revised "Common and Garrison Regulations" in October 1997. Streamlining of military region-level organs began in August 1998.[46] The U.S. Army in the early 1990s managed to cut combat units, but had less success in reducing headquarters staffs and the institutional Army; it will be interesting to see whether the Chinese are any more successful than we were.

Assessment of the Effect of These Reforms.

The most important issue raised by the reform programs is whether they constitute the kernel of a significant power projection capability for the PLA. The short answer, in this analyst's judgement, is no. The logistics reform, if successful, will improve the PLA's ability to sustain its forces in high-intensity combat in or very near the homeland, but it strengthens rather than mitigates their dependence on a geographically fixed logistics infrastructure. Organizations to provide mobile logistics on the battlefield are one precondition for power projection, but the other components required to sustain large maneuver forces in combat outside of China remain absent from Chinese force structure, support concepts, and reform plans. These include logistics elements organic to major maneuver units, making them more self-sufficient in transportation, first- and second-line maintenance, evacuating and repairing battlefield equipment casualties, and maintaining and transporting mobile fuel and supply stockpiles. They include port-opening and airfield-opening packages, capable of creating the far end of a logistical flow structure with minimal reliance on local infrastructure. They include vastly greater airlift and sealift capabilities than the PLA has chosen to buy, and the combatant escorts and refueling units to get the lift to its destination. They include construction engineers to build the land-based transportation infrastructure at the far end of the operation.

The logistics reform program addresses critical weaknesses in the PLA's operational capabilities. If successful, it will improve the PLA's ability to fight a modern high-tech conflict on or near Chinese territory. But it does not contribute significantly to regional, much less global, power projection.

At the operational level, the restructuring of the ground forces and the development of improved doctrine continue the PLA's progress toward competent combined-arms operations. The PLA still has a long way to go toward effective joint service operations, but the new combat regulations firmly establish joint operations as a goal for continued doctrinal development.

More broadly, the reforms represent a major step forward in the regularization of PLA operations. The "first-generation" combat regulations of the 1980s were far too general in their scope to constitute operational doctrine in the Western sense. If the 1999 ordinances prove to be as detailed and comprehensive as initial reporting indicates, they, together with the comprehensive restructuring of the force and the 10-year logistics reform program, will represent a major milestone in the professionalization of the PLA and its slow maturation as a military institution.

CHAPTER 3 - ENDNOTES

1. This term includes all Services of the Chinese armed forces.

2. "Jiang Signs Order on Military Establishment Regulations", *Xinhua*, June 24, 1999, *Foreign Broadcast Information Service* (*FBIS*), SK2506020299.

3. Li Xin and Zhang Dongbo, "Wang Ke Calls for Standardization of Logistics Work," *Xinhua*, November 4, 1998, *FBIS* OW0204103499; Wang Ke, "Adapt to Market Economy, Adapt to Future Combat," *Quishi* (Beijing), No. 23, December 1, 1998, pp. 24-28, *FBIS* OW2312083198; Pei Fang, "Major Operation To Be Performed on Military Logistical System", *Kuang Chiao Ching (Wide Angle)*, Hong Kong, No. 318, March 16, 1999, pp. 50-52, *FBIS* OW0204103499; Commentator, "Unswervingly Implement Central Military Commission's Policy Decision," *Jiefangjun Bao (Liberation Army Daily)*, January 26, 1999, p. 1, *FBIS* OW0302120499; Zhang Dongbo, "Wang Ke, At the Expanded Meeting of Party Committee of the General Logistics Department, Emphasized Doing a Good Job in Reforming Army Logistics," *Xinhua* Domestic Service, December 30, 1998, *FBIS* SK0901031699; *People's Navy*, January 10, 1999; *Rocket Force News*, April 2, 1999; *Renmin Qianxian*, Nanjing MR, April 13, 1999. Unless otherwise indicated, all citations to Chinese military publications are excerpted from various editions of the monthly "PLA Activities Reports" complied by SEROLD Hawaii Inc.

4. Pei Fang; *Rocket Force News*, April 2, 1999; *Renmin Qianxian*, April 13, 1999; Chen Hui, "PRC Armed Forces Carry Out Three Major Reforms," *Xinhua*, June 14, 1999, *FBIS* OW1606021799.

5. Wang Ke, note 2.

6. Wang Ke; *Jiefangjun Bao*, March 22, 1998.

7. Chen Hui, note 6.

8. Wang Ke.

9. Jiang Jiesheng, "Change effected in strategic logistics support," *Jiefangjun Bao*, February 9, 1999, p. 6, *FBIS* OW2402103499; *Directory of People's Republic of China Military Personalities, June 1998*, Honolulu, HI: SEROLD Hawaii Inc., 1998, p. 26.

10. Wang Ke.

11. Chen Hui; *Rocket Force News*, April 2, 1999.

12. Wang Ke.

13. Pei Fang.

14. Chen Hui.

15. Chen Hui.

16. Pei Fang.

17. Li Zuozhi, "Comprehensively Promote Socialization of General Equipment Supply", *Jiefangjun Bao* (*Liberation Army Daily*) February 9, 1999, p. 6, *FBIS* OW0503001799.

18. "PRC Military Moving Logistics To Civilian Sector," *Xinhua* Beijing English service, April 2, 1999, *FBIS* OW2204082199.

19. Luo Yuwen, "Zhang Wannian on All-Army Farm Production," *Xinhua*, April 13, 1999, *FBIS* OW2004104299. For an excellent discussion of the self-sufficiency efforts of Red Army 359th Brigade in Nanniwan, 1940-45, see James Mulvenon, *Soldiers of Fortune: The Rise of the Military-Business Complex in the Chinese People's Liberation Army, 1978-98*, Armonk, NY: M.E. Sharpe, forthcoming 2000.

20. Pei Fang.

21. Wang Zhilei and Zhang Haiping, "Reform and Opening Promote Support Might—A Summary of PLA Logistics Development," *Jiefangjun Bao*, December 20, 1998, pp. 1-2, *FBIS* OW2601053299.

22. Chen Hui.

23. *Zhanshi Bao* (Guangzhou MR), May 15, 1998.

24. *Zhanyou Bao* (Beijing MR), April 24, 1999; *Jiefangjun Bao*, June 6, 1999.

25. *Jiefangjun Bao*, June 29, 1998; *People's Navy*, April 13, 1999.

26. *Air Force News*, April 3, 1999.

27. Wang Ke.

28. Nicholas R. Lardy, *China's Unfinished Economic Revolution*, Washington, DC: Brookings Institution, 1998, pp. 11-12.

29. Various PLA officers in personal communication with the author.

30. Wang Ke.

31. Lardy.

32. Xu Yang, "Jiang Zemin, Zhu Rongji Issue New Army Regulation," *China Daily* (Internet version), July 13, 1999, *FBIS* OW1307033099.

33. Pei Fang.

34. *Rocket Force News*, April 2, 1999.

35. *Qianwei Bao* (Jinan MR), April 7, 1999; *Zhanyou Bao* (Beijing MR), February 10, 1998; *Jiefangjun Bao (Liberation Army Daily)*, July 29, 1998; *Renmin Jundui* (People's Military)(Lanzhou MR), April 8, 1999; *PAP News*, August 4, 1998; *Jiefangjun Bao*, June 6, 1999; *Zhanshi Bao* (Guangzhou MR), May 11, 1998; *Jiefangjun Bao*, June 8, 1998; *Jiefangjun Bao*, June 26, 1998.

36. *Jiefangjun Bao, (Liberation Army Daily)*, June 13, 1999; *Jiefangjun Bao*, June 5, 1998.

37. *Guo Fang*, June 1999; *Chinese Militia*, June 1999; *China National Defense News*, May 3, 1999.

38. Lonnie Henley, "Officer Education in the Chinese PLA," *Problems of Communism*, May-June 1987, pp. 55-71.

39. Editorial, "Basic Guideline for Our Army's Combat Drill in the New Period—Written on the Promulgation of Operational Ordinance of a New Generation," *Jiefangjun Bao*, January 25, 1999, *FBIS* OW0102104099.

40. Interview With General Chief of Staff Fu Quanyou, "Earnestly Implement Operation Decrees and Continue To Enhance Capacity To

'Win Wars'," *Jiefangjun Bao* (*Liberation Army Daily*), February 25, 1999, *FBIS* OW1903042299; Chang Hung, "To Fight a Modern Local War, the PLA Joint Services Put New Operating Regulations Into Effect," *Kuang Chiao Ching*,(*Wide Angle*), Hong Kong, May 16, 1999, No. 320, pp. 46-47, *FBIS* OW0406004299.

41. Ren Xiangdong, "PLA Ground, Naval, and Air Units Implement New-Generation Combat Regulations," *Liaowang,* (Viewpoint) (Beijing), June 7, 1999, *FBIS* OW290608521999.

42. *Jiefangjun Bao* (*Liberation Army Daily*), March 22, 1998; *Qianwei Bao* (Jinan MR), April 10, 1998.

43. *Zhanshi Bao* (Guangzhou MR), January 23, 1998, May 1, 1998; *Jiefangjun Bao* (*Liberation Army Daily*), June 13, 1998.

44. *Zhanyou Bao* (Beijing MR), December 18, 1997.

45. "Country Briefing: People's Republic of China," *Jane's Defence Weekly*, Vol. 30, No. 24, December 16, 1998; Commentator, "Deepen the Study of the Basic Experience of a Certain Artillery Regiment," *Jiefangjun Bao* (*Liberation Army Daily*), January 26, 1999, *FBIS* OW0302121499; Yang Zhenwu, *et al.*, "Jiang Visits Troops in Inner Mongolia," *Xinhua*, February 3, 1999, *FBIS* OW0502014299; *Jiefangjun Bao*, September 11, 1998; Dennis J. Blasko, interviews with PLA officers 1998-99, conversation with author, July 1999.

46. *Jiefangjun Bao* (*Liberation Army Daily*), February 6, 1998; *Zhanyou Bao* (Beijing MR), August 22, 1998.

CHAPTER 4

POTENTIAL APPLICATIONS OF PLA INFORMATION WARFARE CAPABILITIES TO CRITICAL INFRASTRUCTURES

William C. Triplett II

Introduction.

There is a consensus among specialists that the United States is the world leader in Information Warfare (IW).[1] Nevertheless, if the United States were to have a significant *offensive* IW targeting the *critical civilian infrastructure* of foreign countries, details of such a program would not likely be public information. In fact, the phrase, "computer network attack" (CNA) was classified until about 2 years ago.[2] The financial resources devoted to such a hypothetical program, as well as some of the locations[3] where the most sensitive offensive IW Research and Development (R&D) or operations could be carried out, would similarly be restricted information. The capabilities of such a hypothetical program would be closely held to the point of approximating stealth 2 decades ago.[4]

An extensive look at the U.S. official unclassified publications on IW indicates they are mostly devoted to military-on-military applications [Command and Control Warfare], particularly within the Revolution in Military Affairs (RMA) context. There is also a heavy emphasis on *defensive* measures and only a rare mention of U.S. *offensive* capabilities. Roger Molander of RAND candidly outlines one aspect of the problem:

> The sensitivities of our friends and allies and the political-military capital that might accrue to possible

adversaries from increasingly open emphasis on U.S. offensive SIW [Strategic Information Warfare] initiatives has largely kept the more definitive information on these capabilities in the back room. Some SIW offensive capability clearly exists (and some has been demonstrated) but the full potential is politically and militarily sensitive.[5]

Added to the issue of a possible propaganda black eye is the question of countermeasures. There is no physical defense against a nuclear-armed missile except a preemptive strike,[6] something of a high-risk operation. As a result, the declared nuclear powers have adopted a policy of highly publicized and credible deterrent.[7] IW, on the other hand, is subject to effective countermeasures to the point that Electronic Warfare,[8] for example, is subject to a constant offense/defense battle of counter-counter-counter, etc., measures. If it were known, for example, that the United States was successfully targeting foreign power grids for CNA, then a prospective opponent should be able to invest the resources necessary to frustrate that capability.[9]

In short, simply because a certain military capability is not announced does not necessarily mean it does not exist.[10]

Definitions and Scope.

Definitions. IW is in a state of becoming, and definitions are not universally agreed upon.[11] In some cases there is an aversion to using the word "warfare" so that "Information Operations" is the politically correct term of art. Additionally, IW or IO can have an expansive meaning that even includes Civil Affairs. For the purpose of this exercise, IW will principally mean CNA with the inclusion of Precision Strike, Denial and Deception, Special Operations and Psychological Operations, as appropriate.[12]

Scope. The object of warfare is to make an opponent accept your goals, not his. Your goals may be territorial conquest, hegemony without territorial absorption, or even deterrence of offensive operations. For example, one may wish to exercise deterrence against a third party while engaging in operations against a second opponent. Because IW is relatively cheap, it lends itself to sudden assault. Because modern societies are increasingly vulnerable, it could be an effective military tool for reaching your goals.[13]

This exercise is designed to look rather narrowly at how the Chinese People's Liberation Army (PLA) IW capabilities might be applied against foreign critical infrastructures for political/military purposes. A broader study would consider other PRC institutions with IW capability, such as the Ministry of State Security[14] and other IW applications, such as espionage. Such a study is recommended.

A Summary of PLA IW Capabilities and Doctrine.[15]

If the *Liberation Army Daily* (LAD) should announce that the PLA has the capability to change Japan National Railways (JNR) switches at will using CNA from outside of the country, it would cause a sensation. Any responsible Japanese Government would immediately move to head off this obvious threat to public safety. There would probably also be an intense examination in Japan of other parts of the Japanese critical infrastructure of the potential risk. Repercussions would spread to other countries as well. In short order, whatever offensive IW capability the PLA had would be countered both at the JNR and other infrastructure nodes. The Japanese and others in the region would have their alert level raised considerably.

The LAD has not, so far, made such a claim and seems unlikely to do so but arguing from the U.S. analogy, absence of publicity does not necessarily mean absence of capability. From the externalities, however, what can we summarize about the PLA's IW capabilities or even the PLA's *opportunity* to have an IW capability, if it so chooses?

An IW Program. It is fairly well-established that the PLA has some kind of IW program. The U.S. Department of Defense writes, "the PLA has shown an exceptional interest in information warfare."[16] Others rank the PLA program right behind the United States and Russia.[17]

Leadership Support. In a 1998 think piece on future warfare, PLA Chief of Staff General Fu Quanyou added "information warfare" to a list of new forms of combat that also included air strike, missile and electronic warfare.[18] At a Commission on Science, Technology, and Industry for National Defense (COSTIND) meeting in 1995 General Liu Huaqing is alleged to have said:

> Information warfare and electronic warfare are of key importance, while fighting on the ground can only serve to exploit the victory. Hence, China is more convinced (than ever) that as far as the PLA is concerned, a military revolution with information warfare as the core has reached the stage where efforts must be made to catch up with and overtake rivals.[19]

There is some indication that IW has the support of rocket scientist-emeritus Qian Xuesen,[20] and, according to one observer, there is "a push toward establishment of a high level leading group" on IW.[21] One also has to assume that having repeated access to the pages of the LAD indicates some measure of political support for IW, since the LAD is published by the PLA's General Political Department.

Financial Resources. There is no unclassified answer to how much money the PLA is prepared spend on IW. According to one unconfirmed report, the PRC is intending to pour "billions" of Australian-dollar equivalents into the PLA's IW program.[22]

However, it would seem that there is an opportunity to fund an IW program if they wish. At the end of 1999, the PRC claimed approximately 155 billion U.S. dollars in foreign exchange reserves. Billions of U.S. dollars have already gone into a series of very expensive military purchases from Russia, Israel, and other sources. In the spring of 1999, the Republic of China on Taiwan (henceforth Taiwan) found that the PLA had re-created an exact, one-for-one, duplicate of CCK Air Base in central Taiwan. CCK was designed in the 1950s by the U.S. Strategic Air Command with very long runways and ultra-wide taxiways to handle B-52s. Even considering the relatively low cost for construction labor in China and the location in the desert in Gansu Province, creating this duplicate must have been very expensive. It would have required approval at very high levels, and the leadership must have been convinced of its importance. Given how inexpensive a reasonable IW program would look against the re-creation of a SAC base in Western China, it would seem the funds should be available.

IW Theorists. We have a list of the names of the major theorists whose work appears in LAD or similar publications.[23] Whether these are real persons in all cases, or pen names, or committees are unknown.

IW Research and Development Centers. There is a list of IW R&D Centers to which we must add the new Science and Engineering University and the new PLA Information Engineering University at Zhenzhou.[24] However, considering again the American example, some of the more sensitive centers may well be elsewhere.

Tools: SIGINT. The PLA "maintains by far the most extensive signals intelligence (SIGINT) capabilities of all the countries in the Asia/Pacific region."[25] Just this year, according to one unconfirmed report, they expanded their SIGINT facilities to include Cuba, where they may be intercepting telephone calls along the East Coast of the United States.[26]

Tools: Electronics and Telecommunications Infrastructure. Without doubt Beijing is developing the domestic infrastructure that could support a serious IW program. This includes microelectronics, computers, and fiber-optics-based telecommunications.[27]

Tools: High Performance Computers. The Cox Committee noted that between January 1996 and the end of

1998, over 600 American-origin high performance computers (HPCs) were delivered to China. The number is undoubtedly higher today. The Committee believes that such HPCs could prove valuable for offensive IW by giving the PLA the tools to explore "U.S. information networks and their vulnerabilities." Such computers, the Committee says, are also useful for the development of IW associated technologies such as jammers, microwave weapons, and anti-satellite weapons. The Committee notes that the PRC does not permit end-user checks and therefore it cannot be determined with precision where the HPCs are or to what purposes they might be put. There have been a number of recent cases where HPCs made their way to military end-users, creating a sense of uneasiness, at the least.[28]

Tools: Technical Personnel. The Cox-Dicks Committee estimated that "at any given time there are over 100,000 PRC nationals who are either attending U.S. universities or have remained in the United States after graduating from a U.S. university." They noted, correctly, that these students are "a ready target for PRC intelligence officers and PRC Government-controlled organizations, both while they are in the United States and when they return to the PRC."[29] A National Counterintelligence Center report claims the PRC is recruiting ethnic Chinese students around the world who have expertise in certain weapons and weapons-related skills including computers.[30]

As early as the fall of 1994 the PLA was telling the domestic Chinese press that it was recruiting "returned students" who have experience in "supercomputers" and "artificial intelligence," both building blocks for IW.[31] In the summer of 1999, President Jiang announced the establishment of a cash awards fund designed to lure Chinese students with science and technology skills back to China.[32]

There is definitely a domestic training program as well. In July 1999 the LAD discussed the need for "Internet warriors" who could take part in "Internet combat."[33] The

PLA is also offering scholarships for Information Warfare students at Beijing and Qinghua Universities.[34]

Tools: High-powered Microwave Weapons. As the Cox-Dicks Committee noted, high-powered microwave weapons (HPMs), sometimes known as radio frequency weapons, have a particularly important role to play in any serious IW program. The report also accused the PRC of illegally obtaining U.S. R&D information on "electromagnetic weapons technology."[35] This apparently occurred in the late 1990s.

China has a national HPM program as does the United States, Russia, the United Kingdom, and France. China's foremost expert on HPMs is Dr. Lin Weigan (University of California, Berkeley, 1950) and there are institutes devoted to this research.[36]

PLA writings make a clear connection between HPMs and IW. For example, in 1995, a COSTIND researcher praised HPMs' ability to "destroy the opponent's electronic equipment and electronic telecommunications systems. This is a special kind of information-intensified weapon for waging information warfare."[37]

Russia is considered to be the world's leader in HPMs, with research concentrated at the Institute of Applied Physics and Lebedev Physics Institute.[38] Some of the Russian HPMs are suitcase-size and have been sold on the open market to at least Sweden and Australia.[39] If the size could be further reduced, they might be deliverable by Unmanned Aerial Vehicles (UAV). There are no open-source indications that the PLA has purchased Russian HPMs of any size or that Russian HPM specialists are assisting the Chinese program, but under the current circumstances it is certainly worrisome.[40]

Tools: Precision Strike. U.S. doctrine for offensive Information Operations include "physical attack/ destruction," i.e., long-range precision strike.[41] PLA theorists have no difficulty making the link between "precision attacks" and "information warfare." It is abundantly clear that serious PLA resources are going into offensive missile strike capability, and that these efforts are seeing some success.[42]

The connection between IW and air power is also not lost on the PLA, as two authors pointed out in *China Military Science*: "Air power's exclusive strength in information warfare has made it a key factor for armies of all countries."[43]

Tools: Special Operations. U.S. doctrine for Information Operations also includes Special Operations.[44] In recent years there has been considerable public attention paid to PLA special forces/special operations, including airborne troops, perhaps indicating that they are receiving more resources. In the summer of 1999, *Wen Wei Pao* published a report on "Special Forces: Emergency Mobilization for Long-Range Assault." The report discusses the new equipment available to Chinese Special Forces, including motorized hang gliders, and asserts that their mission includes "going deep into the enemy heartland."[45]

Doctrine: Reading American Literature. PLA theorists are clearly reading the U.S. open-source literature.[46] For example, U.S. Army Field Manual 100-6, "Information Operations," has been translated by the PLA's Military Science Publishing House.[47]

As another example, in the summer of 1999 there was an article on a Beijing-based Internet site discussing the June 3, 1995, National Defense University exercise, "The Day After...in Cyberspace," which appears as Appendix C in the RAND publication *Strategic Information Warfare* published in 1996.[48] PLA writers tend to use virtually the same definitions of IW that the United States does: "electronic warfare, tactical deception, strategic deterrence, propaganda warfare, psychological warfare, computer warfare, and command and control warfare."[49] From time to time, there is talk of an IW with Chinese characteristics, but there is no evidence that such a creature exists as yet. PLA theorists speak of IW as a new form of "People's War," meaning civilian technicians in uniform at the keyboard, but that would be the case under any Western program as well.

Doctrine: Writing Their Own Literature. The theorists are certainly talking about it. Articles have appeared in LAD on the subject since 1987, and in the first 9 months of 1999 the words "information warfare" appeared in LAD at least 10 times.[50] They began publishing books on the subject as early 1985, and there are important symposia covering IW sponsored by COSTIND.[51]

Arguably the most significant article appeared in LAD on November 11, 1999. The three authors argued that the PLA should create a "net force" equal to the PLA Army, Navy, and Air Force. They also argued for an *"offensive technology"* capability.

Doctrine: Preemptive Strike. One might argue that the PLA has a long association with surprise attack, i.e., preemptive strike. Certainly the Vietnamese would agree. It is, therefore, not surprising that PLA IW theorists have been quick to see IW as a preemptive strike weapon. In an often-quoted 1996 LAD article, Lu Linzhi pointed out that, "In military affairs, launching a preemptive strike has always been an effective way in which the party at a disadvantage may overpower its stronger opponent." Lu calls for zeroing in on "enemy troops as well as the war-making machine."[52]

Training and Exercises. There has been a lot of discussion going back several years about training for IW. For example, in February 1999 the LAD ran a story titled, "Train Talented People at Different Levels for Information Warfare."[53] According to a COSTIND-controlled publication, the PLA's Communications Command Academy is offering a cross-disciplinary course on IW based on textbooks prepared by experts from the PLA Engineering Institute, the General Staff Department, the Academy of Military Sciences and the National Defense University. The

Annual Harvard Senior Military Officers program for PLA officers includes IW training.[54]

By one account, IW was included in the spring 1996 exercises in the Taiwan Strait.[55] Two years later Shenyang Military Region was boasting of its IW exercises in LAD, "Accept a New Challenge, Exercise in New Topics—More than 20 Achievements Tested by Shenyang Military Region Live Information Warfare Exercise." Just recently Lanzhou Military Region was conducting "Red Team-Blue Team" Information Warfare exercises that opened with a surprise network attack by the "Blue Team."[56]

In summary, the Chinese have an IW program, they have a reasonable level of leadership support, sufficient available resources, the necessary tools (or are acquiring them), and they are working on doctrine in the right direction and conducting military exercises that feature IW prominently. Their level of competence is as yet unknown, perhaps to themselves as well.

The next question is "What's the most likely application?"

Potential Applications I—The United States.

Today there exists a level of concern bordering on alarm in official Washington over U.S. vulnerability to CNA against critical infrastructures. A former Director of NSA called the United States the most vulnerable country in the world to CNA and in 1996 then-CIA Director John Deutch expressed similar concerns to the Senate.[57] The issue is being studied in the public policy institutes. Some two major commissions have convened on the matter, and the Defense Science Board has reported on it. Even the Senate Y2K Committee drew attention to it. A series of useful U.S. General Accounting Office reports on the threats to critical infrastructures have been written.[58]

The Public Policy Institutes. The Center for Strategic and International Studies (CSIS) recently completed a major IW study subtitled *Averting an Electronic Waterloo.* The opening paragraph reads:

> The United States is now exposed to a host of new **threats** to the economy, indeed **to the whole of society**. It has erected immensely complex information systems on insecure foundations. The ability to network has far outpaced the ability to protect networks. The economy is totally dependent on these systems. **America's** adversaries and **enemies** recognize this dependency and are developing **weapons of mass disruption** and destruction.[59][Emphasis added]

Nor is CSIS alone. RAND's *Strategic Information Warfare—A New Face of War* contains the following assessment:

U.S. homeland is vulnerable

- Cyberspace efficiencies-use-dependence-vulnerability cycle especially acute in U.S.

- U.S. info-based infrastructures present lucrative strategic targets.

- IW weapons less physically destructive than Russian ICBMs, but much cheaper to field, and probability of use in conflict is much higher.

You lose U.S. as sanctuary.[60]

The Commissions. In 1997, the President's Commission on Critical Infrastructure Protection released its report, "Critical Foundations—Protecting America's Infrastructures." The Commission was created in response to the Oklahoma City bombing of 1995. It began looking at physical threats but moved fairly quickly into cyber threats. It examined five sectors—information and communications; banking and finance; energy, including electrical power, oil and gas; physical distribution; and vital human services. The Commission was very alarmed at what it found.

1. There is an increasing dependence on critical infrastructures.

2. There is an increasing vulnerability to those infrastructures from a wide range of threats, particularly IW.

3. There is a lack of public awareness to the threats.

4. There is no national focus for protection of critical infrastructures.

As the Commission findings noted,

> Potentially serious cyber attacks can be conceived and planned without detectable logistic preparation. They can be invisibly reconnoitered, clandestinely rehearsed, and then mounted in a matter of minutes or even seconds without revealing the identity and location of the attacker.[61]

In the fall of 1999, the United States Commission on National Security/21st Century released its report, "New World Coming: American Security in the 21st Century." The Commission, chaired by former Senators Gary Hart (D-CO) and Warren Rudman (R-NH), notes that "the United States and its allies are particularly vulnerable" to IW attacks on the critical infrastructure because "the health, welfare and prosperity of the citizens of the developed world depend on this infrastructure." The Commission asked its readers to "Imagine, for example, a well-planned attack against the air traffic control network on the east coast of the United States as more than 200 commercial aircraft are trying to land in rain and fog on any given weekday morning."[62]

The Defense Science Board Report. In the fall of 1996, the Defense Science Board presented its report on threats to DOD systems. The Board concluded,

> Because of its ever-increasing dependence on information and information technology, the United States is one of the most vulnerable nations to information warfare attacks.

Further,

> We conclude that there is a need for extraordinary action to deal with the present and emerging challenges of defending against possible information warfare attacks on facilities, information systems, and networks of the United States which would seriously affect the ability of the Department of Defense to carry about its assigned missions and functions.

Finally, the Board warned that,

> increasing reliance on information systems to operate, maintain and monitor infrastructures . . . creates a tunnel of vulnerability previously unrealized in the history of conflict.[63]

Vulnerabilities. To what extent are the concerns expressed above valid? Consider the following:

On June 16, 1999, the sanitation department of Van Nuys, California ran a Y2K compliance check. At midnight they were testing a backup electrical system when they received a frantic call from a park ranger. Raw, untreated sewage was pouring out of a manhole cover and spilling into a park. Later it was estimated that four million gallons of sewage had been released. A computer had mistakenly closed a gate that should have remained open to control transfer of the sewage. A programmer's error in 1985 seems to have been the culprit. It cost the taxpayers about $100,000 to clean up the mess.[64]

That was one line of code, sending one message to one computer-controlled gate. And it was an accident. A competent, motivated, hostile force might be able to do the following:

- Change the dose levels in prescription medicines at pharmaceutical plants so that thousands of people would be poisoned.[65]

- Infiltrate the manufacturing process for baby food so that the standard components would be increased by 400 percent (to toxic levels).[66]

- In a similar manner, taint the processed food industry for adults: restaurants, hotels, hospitals, and retirement homes.

- Subtly change the radar signals picked up at airports so that air traffic controllers would unknowingly put passenger planes on the same flight path.

- Stage a surprise attack on many of the automated gasoline refineries in the nation, causing enormous, out-of-control fires that would inundate emergency officials and lead to immediate gasoline rationing.[67]

- Contaminate the city water systems, turn the valves backwards at the sewer systems, shut down the electric power grid, overload the natural gas pipeline system.

- Loot the bank accounts of selected persons, transferring all their funds overseas and wiping out records of their existence.[68]

- Attack the identities of selected persons, eliminating their social security records, Veterans Department records, driver's license numbers, bank and credit card numbers and accounts and so on.[69]

All of these examples have either been tried or are of concern to competent officials.

Example: SCADAs and Related Systems. "SCADA" is the acronym for "Supervisory Control and Data Acquisition," the computer-controlled gates which control an increasing percentage of American infrastructures. The SCADA systems are composed of three parts—a master station, one or more remotes, and the custom software that controls them. In a public water system, for example, they would control when the pumps start and stop, how much chlorine is injected into the system, which valves are open and which are shut, and so forth. According to the President's Commission, "Cyber vulnerabilities include the increasing reliance on SCADA systems for the flow and pressure of water supplies."

Related systems called "Distributed Control Systems" are routinely used in food processing, pharmaceutical, and chemical facilities. Other SCADA-related systems control railways and trucking operations.

Consider the electrical grid. It might have thousands of SCADAs or SCADA related computer-controlled switches that are hundreds of miles apart all tied to the central computer. Communications from the center to the remotes is by radio, and typically the radio communications links are not encrypted. The SCADAs may also be maintained by outside contractors whose personnel may or may not have received background checks. According to the President's Commission, the systems rely on "dial-back modems that can be bypassed," which means that an outside team could gain access to them either by capturing and overriding the radio signals to the remotes, or by using the telephone system and the dial-back modems to gain control over the entire network. Once inside, a hostile force could wreck havoc on the system.

Pharmaceutical plants, processed-food factories, wastewater treatment plants, refineries, chemical plants, electrical grids, oil and gas pipeline systems—all are susceptible to manipulation and destruction.[70]

Further Example: The Refineries. The refineries are probably the most vulnerable. As of this writing, the United States has 160 active refineries. Some are singles, but others are clustered, often with associated chemical plants. As the President's Commission noted, "Large refineries (greater than 250,000 barrel capacity) in California, Texas, and Louisiana would be attractive targets for physical or cyber attack."[71]

An attack might look like this: The CNA team operating either abroad or from a clandestine U.S. location would access the system through the dial-back modems. The system would be mapped thoroughly and a "backdoor" would be left in place so it could be monitored easily for any changes in the system. CNAs would be rehearsed until the leadership had a high level of confidence. Using some cover story such as seeking employment or maintenance, a special operations individual or team would gain physical access to the command building. It, too, would be mapped looking for doors, windows, or other entranceways that give a direct line-of-sight from the outside into the area where the central computer and backups are located.

The attack would be in two phases. First, the CNA team would deliver false messages to the SCADAs, setting off cascading fires and explosions. Next, at the proper moment the special operations team would set off an HPM,[72] disintegrating the electronics. This would make it impossible for the refinery's Information Technology team to recover, isolate the fires, or regain control over the system. If a number of refineries and chemical plants in the same area were attacked simultaneously, the effects would be catastrophic—the emergency response teams would be overwhelmed, thousands of people would be killed or injured, and gasoline rationing would be at World War II levels for months.

Potential Applications II.

Taiwan. In 1995, a decade after the PLA started on IW, Taiwan's military establishment was complacent: "As far as our side is concerned, the defense of the Taiwan Strait area is mainly a matter of resisting landing operations."[73] As late as the fall of 1998, two well-respected researchers from Taiwan would write that Taiwan's military restructuring program is "not designed to counter possible electronic and information warfare waged by the PLA."[74]

All that is different now. At a March 1999 conference, Defense Minister Tang Fei surprised his audience by claiming that Taiwan may be powerless against an IW attack from the mainland. Chief of the General Staff Tang

Yaoming told a seminar, "A massive information attack may lead to chaos here." Dr. Lin Chong-pin of the Mainland Affairs Council has gone out with a public warning of Taiwan's increasing vulnerability to an IW attack. From ignoring the issue, Taipei's newspapers have been full of articles about IW, particularly as political figures have begun to take notice.[75]

The response is a bit late, but it is on the mark. Taiwan, like the United States, is an industrial society tied more and more to computer systems. Whatever applies to American infrastructure vulnerabilities applies to Taiwan as well. The five sectors examined by the American President's Commission—finance, telecommunications, energy, physical distribution, and public services—have the same risk in Taipei as they do in Washington.

Taipei may have some additional exposure that Washington has escaped. PLA theorist Wang Baocun talks about the idea of "virtual warfare" where:

> One is using virtual reality and computer imaging technology during the course of war to construct images of the enemy's supreme commander and have him issue statements that are not conducive to the conduct of the war, such as having him declare though his homeland's television system that, for various reasons, hostilities with the enemy have ceased and all military forces are being withdrawn.[76]

Considering how often President Lee has appeared on television, and how relatively easy it would be to capture his likeness and manipulate it, Taipei should take warning.

Taipei has other concerns as well. In the event of war, the early warning network would be the first to be attacked. It is unlikely that every radar on Taiwan would be taken out in the first wave. If the surviving radar shows an attack coming from the south, is this a true image, or a "virtual fleet" created by Colonel Wang and his comrades? Could the real attack be coming from the north? If your troops know that the Other Side has been inside your system, even once, how much trust will they place in orders coming from Central Command, particularly maneuver orders that would require them to leave a prepared and presumably secure position?

Taipei's vulnerabilities to IW are only limited by human imagination. Clean out the bank accounts of the political and economic elite and transfer the proceeds overseas? Why not let them have their money and then publicize the fact that they are running for cover? Close down all the civilian airports? Why not let people think one is open in order to increase the panic and chaos as the elite tries to reach presumed safety? As Dr. Lin pointed out this year, "Our psychological defense is almost nil."[77]

Japan. Raisuke Miyawaki, senior advisor to the Japanese Commission on Critical Infrastructure Protection, told a CSIS conference in June 1999, "Key elements of Japan's information infrastructure remain fragile and open to cyberattack." He claims there is a "leadership void" in Japan because of "a lack of technology understanding" at the top. Japan is only going through the motions of infrastructure protection, he says, because the United States is pressuring it to do something about the problem.[78]

Whereas the refineries are probably the most lucrative U.S. target, the railways and subways are probably the most problematic in Japan.[79] Japanese hackers have already successfully attacked the computer control systems of commuter trains, paralyzing major cities for hours.[80] The 1995 NDU exercise considered a "new high-speed Metro-Superliner traveling at 300 km/hr" slams into a misrouted freight train somewhere in Maryland.[81] The team estimated 60 deaths, but in the Japanese context, it would be much higher. A surprise attack on the Japanese national railways would aim at multiple accidents during the height of the rush hour, followed by a total shutdown of the emergency response team. Those who did not perish immediately from the shock of the collisions would die lingering, painful deaths because no one could care for them.

Instrument of Choice.

If there is to be armed conflict in the Strait, the PLA's instrument of choice could be IW applied to Taiwan, the United States, and Japan. Taiwan would receive the full brunt of the campaign: preemptive precision strikes, massive CNAs, aggressive deception, and psychological warfare on a grand and imaginative scale. If a well-known political figure could be coopted, one option might be to use Special Forces to take over a military facility, install him or her there, and rescind any calls for outside assistance.

Since the aim would be to keep the United States and Japan out of the fray, CNA could be applied here at a reduced level as part of a diplomatic campaign. However, the threat of major destruction to critical infrastructures would always hover in the background.

The primary attractiveness of such a program would be power projection. As the RAND report demonstrates, there is no other way to reach the U.S. homeland short of nuclear Armageddon. The U.S. mainland has not seen war on the home turf since Appomattox, and no one could determine the psychological effects a serious attack would produce.[82] On Taiwan and Japan, probably no one under 60 has any real memory of World War II.

At the same time, it is the perfect match of vulnerability against capability. Responsible officials in the United States, Taiwan, and Japan are practically announcing to the world how at-risk their most critical infrastructures are. At the same time, as industrial societies become every day more and more interconnected, they become even more vulnerable.

Used cleverly, IW can be a very useful diplomatic tool because it is the consummate political/military rheostat. In the U.S. context, an opponent could go from causing minor inconvenience by turning all the stop-signals of a major city to blinking red, or creating massive civilian casualties, and every gradation in between.

There can be no question that PLA theorists are aware of the opportunity IW presents to the Taiwan context. The very RAND publication that includes the information above on the vulnerability of the U.S. homeland was discussed in an LAD article in 1998[83] and an Internet article in June 1999.[84] The RAND publications are all available online, as is the CSIS report. "Critical Foundations" also can be downloaded from the Internet. From a straight military point of view, the PLA is unlikely, it seems, to make the mistakes Britain and France did in allowing military innovation to pass them by during the interwar period.[85]

Responses.

The Administration. While the GAO, RAND, and others have been pointing to U.S. vulnerabilities to IW for years, the U.S. Government has been slow off the mark. In May 1998 President Clinton issued Presidential Decision Directive (PDD) 63 "Critical Infrastructure Protection," which set goals and established administrative mechanisms. In keeping with many government projects, the program is over budget and behind schedule as of this writing.

Congress. Congress has been equally slow in responding, but it has a very large trump card. Saying that IW is the PLA's best option in the Taiwan context does not mean it is without risk. The real danger to the PRC would be escalation led by Congress. Since 1985, Congress has imposed economic sanctions on South Africa, Iraq, Iran, Libya, Cuba, and Burma. It threw Toshiba Machine out of the United States for 3 years and ultimately cost the parent corporation half a billion dollars in lost business. Congress also passed sanctions on foreign corporations who export nuclear technology, chemical or biological weapons technology, missiles, or advanced conventional weapons to certain countries. With the exception of sanctions on Iraq and Cuba, all of these actions were taken over the strenuous objections of the Executive Branch.

Given congressional hostility to Beijing now, any war move toward Taiwan is very likely to trigger a serious response. Congress, if it so chooses, has great leverage up to and including a full economic embargo against the PRC, and confiscation of all property held in the United States by PRC citizens. As it demonstrated in the South Africa case, Congress is not afraid of engaging in a secondary boycott and, given the size of the U.S. economy, can throw considerable weight around. It is worth pointing out that a Democrat-controlled Congress with a veto-proof margin passed the Taiwan Relations Act while a Democrat was in the White House. By the same token, the South Africa sanctions legislation was drafted by a Republican Senate and passed over a Republican President's veto.[86] If democratic Taiwan passes to a communist regime by force and violence (with substantial civilian casualties), Congress can turn the lights out, and there will be little the White House can do about it.[87]

Beijing is clearly aware of this. In a recent interview Senior Colonel Wang Baoqing of the Academy of Military Sciences listed "economic sanctions against the mainland" as his first estimate of U.S. response in the event of an attempt by the PRC to take over Taiwan.[88]

In conclusion, the United States, the PRC, Taiwan, and even Japan are entering a dangerous period. The gap between what offensive IW is capable of accomplishing and the nonaction by the defenders grows every day. In fact, the defenders are advertising their vulnerability in hope of generating public support to do something about it. As the RAND study notes, "the probability of use [of IW] in conflict is much higher" than the nuclear exchange which formerly dominated official concerns.[89] For those with the capability to use IW against critical infrastructures, the temptation is to apply it.

99

CHAPTER 4 - ENDNOTES

1. For example, see *The Changing Role of Information Warfare*, Zalmay M. Khalilzad and John P. White, eds., Santa Monica: RAND, 1999, p. 275.

2. Guidance concerning CNA in *Joint Publication 3-13*, "Joint Doctrine for Information Operations," leads to Annex A, which is classified. See pp. II5-6.

3. On October 1, 1999, the U.S. Space Command took public responsibility for U.S. IW activities. See "Pentagon Sets Up New Center for Waging Cyberwarfare," *The New York Times*, October 8, 1999. However, this does not preclude other nonpublic sites of significance in this field.

4. This does not mean that there is no public information on Federal Government IW plans and programs, certainly in the broad sense. A surprising amount of information is freely available on the Internet. For example, one can download *Joint Publication 3-05*, "Doctrine for Joint Special Operations," *Joint Publication 3-13*, "Joint Doctrine for Information Operations," *Joint Publication 3-13.1*, "Joint Doctrine for Command and Control Warfare," *Joint Publication 3-58*, "Joint Doctrine for Military Deception," and *U.S. Army Field Manual 100-6*, "Information Operations." The National Defense University website has a number of useful documents, and the reports of various governmental commissions are mostly available. Those reports that are classified often have an extensive unclassified annex where the specialist can glean important insights. Outside of official U.S. Government publications, in recent years there have been a number of books, including several by RAND, which seem to be based, at least in part, on U.S. Government sources. The popular press has also produced a number of books, the most widely known of which is James Adams' *The Next World War*, which may have been inspired by unofficial U.S. Government assistance.

5. Roger Molander, *et al.*, *Strategic Information Warfare Rising*, Santa Monica: RAND, 1998, p. 27.

6. This is called "Attacking the archer, not the arrow." For example, an SSBN could be sunk before it can launch its missile load, but once the missiles have cleared the surface of the water, the only hope for the defense is a major malfunction.

7. i.e., Mutually Assured Destruction.

8. In U.S. doctrine, EW is part of Information Operations, and CNA is contained in EW. See *Joint Publication 3-13*.

9. For example, the United States has been very reluctant to discuss American IW operations in Kosovo other than to acknowledge that they occurred. See "US Opened Cyber-War During Kosovo Fight," Scripps-Howard, in *The Washington Times*, October 25, 1999.

10. When Senator Carl Levin (D-Mich) asked CIA Director George Tenet at a closed Senate meeting whether the United States had a significant offensive IW capability, Tenet is alleged to have answered cryptically, "We are not asleep at the switch in this regard." *The Jerusalem Post*, May 20, 1999.

11. Group Captain Peter Layton of the Air Power Studies Centre (Australia) uses the term "Network-Centric Warfare." See APSC Paper Number 74 for 1999.

12. *Joint Publication 3-13*.

13. A useful study is Ryan Henry and C. Edward Peartree, "Military Theory and Information Warfare," *Parameters*, Autumn 1998, page 121. Also recommended is Edward Waltz, "The US Transition to Information Warfare," *The Journal of Electronic Defense*, December 1998, p. 35.

14. See *Reuters*, December 16, 1999, for a complaint by a Hong Kong-based human rights organization that the MSS was using IW as a weapon.

15. The following summary of PLA IW capabilities and doctrine draws heavily from U.S. Air Force Major Mark Stokes, *China's Strategic Modernization: Implications for the United States*, Carlisle: U.S. Army War College, Strategic Studies Institute, 1999 (hereafter *CSM*); Dr. James Mulvenon, "The PLA and Information Warfare" in CAPS/RAND Conference Proceedings, 1998; and "US National Security and Military/Commercial Concerns with the People's Republic of China" (hereafter Cox-Dicks report).

16. Report to Congress Pursuant to section 1226 of the FY98 National Defense Authorization Act.

17. Mulvenon, p. 175.

18. *Qiushi*, March 16, 1998, p. 2-6. Fu is also alleged to have endorsed a new book on IW.

19. "Latest Trends in China's Military Revolution," in *Hong Kong Hsin Pao,* In Chinese, February 9, 1996, p.17, *FBIS-CHI*-96-047.

20. Stokes, *CSM*, p. 16.

21. *Ibid.*, p. 27.

22. "China Gears Army for Cyber-War," *The Australian,* November 10, 1999.

23. See Mulvenon, Table 1.

24. See *Ibid.*, Table 2; *Xinhau,* domestic service, in Chinese, July 2, 1999; *Xinhua* in English, November 17, 1999, respectively.

25. Desmond Ball, *Jane's Intelligence Review,* August 1995. Stokes, *CSM*, in agreement, p. 20.

26. *El Nuevo Herald,* Miami, in Spanish, June 24, 1999.

27. See Stokes, *CSM*, especially Chapters II and III. See also *Xinhua* announcement of December 28, 1999, on the rapid growth of the "information industry" in 1999.

28. AUH, "U.S. National Security and Military/Commercial Concerns with the People's Republic of China," Vol. 1, May 1999, pp. xxx, 116, 134, 137-140.

29. *Ibid.*, p. 41.

30. *The Washington Times,* October 11, 1999.

31. *Xinhua,* domestic service in Chinese, October 5, 1994.

32. *The Hong Kong Standard,* August 24, 1999.

33. Shen Weili, "Stressing the Study of Internet Combat," *Liberation Army Daily,* July 27, 1999, p. 6.

34. *The Hong Kong Standard,* November 19, 1999.

35. Cox, Vol. I, p. xiii. It is not clear from the reference whether this refers to HPMs or electromagnetic pulse devices. Both have possible ABM and AST applications. HPMs and EMPs tend to be related in the R&D arena. See Stokes, *CSM*, p. 103. See also Stanley T. Hosmer, "The Information Revolution and Psychological Effects," in *The Changing Role of Information in Warfare*, Khalilzad and White, eds., footnote 223.

36. Stokes, *CSM*, pp. 102, 104.

37. *Guoji Hangkong,* in *FBIS-CHI*-95-114, March 5, 1995.

38. Stokes, *CSM*, p. 105.

39. *Sevenska Dagblat*, January 21, 1998.

40. Chinese officials have, however, attended arms shows where Russian firms were demonstrating HPMs. See *Nezavisimoe Voennoe Obozrenie*, No. 23, 1998, p. 6.

41. *Joint Publication 3-13*, p. viii.

42. *Jiefangjun Bao*, August 10, 1999, p. 6; Mark Stokes, Chapter 5, this volume.

43. *Zhongguo Junshi Kexue*, February 20, 1996.

44. *Joint Publication 3-13*, p. viii. *Joint Publication 3-05*, "Doctrine for Joint Special Operations," April 17, 1998, p. II-4, lists "Information Operations" as one of "Special Operations Principal Missions."

45. *Liberation Army Daily,* September 25, 1996; *Wen Wei Po*, July 15, 1999, p. A5.

46. Mulvenon, p. 176.

47. Noted in Political Commissars Qiao Liang and Wang Xiangsui, *Unrestricted Warfare, FBIS,* trans., Chapter 1.

48. *Beijing Jianchuan Zhishi*, in Chinese, June 30, 1999.

49. *Jiefangjun Bao*, June 11, 1996.

50. *Ibid.*; *Ibid.*, February 2, 1999.

51. *Ibid.*; Stokes, *CSM*, p. 15; Mulvenon, footnote 3.

52. *Jiefangjun Bao*, February 14, 1996.

53. *Ibid.*, February 2, 1999, p. 6.

54. *Beijing Keji Ribao*, April 27, 1999; *Xinhua,* in English, June 17, 1998.

55. *Kuang Chiao Ching*, in Chinese, April 16, 1996.

56. *Jiefangjun Bao*, March 30, 1998; *Ibid.*, July 22, 1999, pp. 1, 2.

57. Statement by Admiral John M. McConnell, Director of the National Security Agency, 1995, quoted in John Fialka, *War by Other Means,* Norton, 1997, p. 105; *De Morgen*, in Dutch, Brussels, June 27, 1996.

58. Senator Robert F. Bennett, Chairman, "Investigating the Year 2000 Problem: The 100 Day Report," the U.S. Senate Special Committee on the Year 2000 Technology Problem, p. 191. For a list through October 1999, see Appendix II of the GAO October 1999 report, "Critical Infrastructure Protection," GAO/AIMD-00-01.

59. The CSIS study was supported by a number of government agencies, Capitol Hill, the Intelligence Community, law enforcement, the national laboratories, national security contractors, and two major universities; "Cybercrime . . . Cyberterrorism . . . Cyberwarfare," CSIS, 1998, p. xii.

60. *Strategic Information Warfare—A New Face of War*, Santa Monica: RAND, 1996, p. 33.

61. "Critical Foundations—Protecting America's Infrastructures," President's Commission on Critical Infrastructure Protection Report Summary, Washington, DC, October 1997, p. 6.

62. "New World Coming: American Security in the 21st Century," Report Supporting Research and Analysis, United States Commission on National Security/21st Century, 1999, p. 52.

63. "Report of the Defense Science Board Task Force on Information Warfare-Defense," Report presented to Undersecretary of Defense for Acquisition and Technology, November 1996.

64. *Los Angeles Times,* June 18, 1999.

65. Russian specialists have raised this concern. See *Komsomolskaya Pravda*, Moscow, in Russian, December 16, 1998, p. 5.

66. This has already been done by someone who broke in through a vendor's computer system that was integrated into the baby food manufacturer's network. Fortunately, the manufacturer found out in time. *Christian Science Monitor*, June 24, 1999.

67. Modern gasoline refineries are controlled by a very few employees opening and shutting valves by computer. Russian

specialists have also raised this possibility. See *Komsomolskaya Pravda*.

68. In 1995 a Russian biochemistry student broke into a major New York bank's computer and transferred $12 million overseas. *Parameters*, Winter 1996-97, pp. 81-91.

69. A frightening amount of personal information relating to person's identity is available on the Internet. Armed with the various account numbers, a hostile force should be able to wreck havoc inside a major computer system.

70. *Beijing Jianchuan Zhishi* in Chinese, June 30, 1999, pp. 13, A27.

71. *Ibid.*, repeated warning in Chapter III as well.

72. Possibly by using a UAV.

73. *Taipei Fengweita Taiwan*, November 1, 1995, pp. 342-351.

74. See Arthur Shu-fan Ding and Alexander Chieh-cheng Huang, "Taiwan's Military in the 21st Century: Redefinition and Reorganization" in Larry M. Wortzel, ed., *The Chinese Armed Forces in the 21st Century*, Carlisle: U.S. Army War College, Strategic Studies Institute, December 1999, p. 282.

75. *Taipei Hsin Hsin Wen*, in Chinese, March 18, 1995; *China Times*, May 13, 1999; Taiwan, Central News Agency, May 31, 1999, respectively.

76. *Beijing Zhongguo Junshi Kexue*, No. 4, in Chinese, November 20, 1997, p. 102.

77. *Taipei Hsin Hsin Wen,* in Chinese, March 31, 1999 *(FBIS)*.

78. Mr. Miyawaki is a senior Japanese civil servant who has served as press spokesman for at least one Prime Minister.

79. A 1998 RAND report prepared for the Office of the Secretary of Defense (Molander, *et al.*, *Strategic Information Warfare Rising*) has a Taiwan Strait scenario and notes the Japanese rail systems as particularly vulnerable, p. 53.

80. Matthew Devost, Brian Houghton and Neal Pollard, "Information Terrorism: Can You Trust Your Toaster?," in *Sun Tzu and the Art of War in Information Warfare*, Bob Neilson, ed., Paper prepared

for the National Defense University, available on the NDU website at *http://www.ndu.edu/inss/siws/ch3.html*.

81. *Strategic Information Warfare*, p. 65.

82. The only other hostile action against American territory was Pearl Harbor for about 2 hours, and the takeover of some Aleutian islands for a year.

83. *Jiefangjun Bao*, July 7, 1998, p. 6.

84. *Beijing Jianchuan Zhishi*, in Chinese, June 30, 1999.

85. See, generally, Williamson Murry and Allan R. Millett, eds., *Military Innovation in the Interwar Period*, Cambridge, 1996. For an indication of PLA understanding of military innovation, see *Jiefangjun Bao*, July 7, 1998, p. 6.

86. The South Africa Sanctions legislation of 1986 was produced by the Senate Foreign Relations Committee chaired by Sen. Richard Lugar [R-Ind] and was accepted verbatim by the House.

87. There is no question that China's economy is much larger than Iraq or Libya, and taking China out of the world economy would be very painful.

88. *Zhongguo Xinshe*, September 7, 1999.

89. Lieutenant General William Donahue, Commander of the Air Force Communications and Information Center, told a fall 1999 conference that hackers had attempted to "take down NATO networks" during the recent military operation in Kosovo. A number of these hackers were traced back to Chinese Internet addresses although he declined to indicate whether they came from government-controlled networks.

CHAPTER 5

CHINA'S MILITARY SPACE AND CONVENTIONAL THEATER MISSILE DEVELOPMENT: IMPLICATIONS FOR SECURITY IN THE TAIWAN STRAIT

Mark A. Stokes

INTRODUCTION

The People's Republic of China (PRC) is developing one of the most daunting conventional theater missile challenges in the world. Theater missiles and supporting space assets are emerging as one of the most important political and operational tools of the People's Liberation Army (PLA). A large arsenal of highly accurate and lethal theater missiles serves as a "trump card" (*shashoujian*), a revolutionary departure from the PLA of the past. The PLA's theater missiles and a supporting space-based surveillance network are emerging not only as a tool of psychological warfare but as a potentially devastating weapon of military utility.[1]

Theater ballistic and land attack cruise missiles, supported by space-based reconnaissance, appear likely to emerge as a cornerstone of PLA warfighting early in the 21st century. A growing sector of the PLA believes strategic attack through theater missile strikes is the best way to even out the playing field when fighting against a technologically superior force. The concept of strategic attack involves pitting one's strengths against an enemy's weakness, waging an asymmetrical strategy using overwhelming offensive capabilities. Theater missiles offer a lethal means of striking targets that most directly relate to the opponent's ability to sustain operations. According to PLA National Defense University officials, "the guiding strategic principle of China's new era military is active defense (*jiji fangyu*), of which the required essence is offensive operations against theater targets."[2]

Beijing's move toward a force dominated by offensive theater missiles could have significant implications for regional stability, creating the first theater in which highly accurate *conventional* ballistic missiles dominate the strategic landscape. Oddly enough, PRC officials, echoed by many within U.S. Government and academic circles, argue that defenses against the growing PLA conventional theater missile threat would be destabilizing since they would spark an arms race. The misplaced focus on missile defenses within academic and policy communities in the United States has resulted in neglect of the dangers presented by Beijing's growing arsenal of increasingly accurate and lethal theater missiles. This arsenal, together with a preemptive doctrine, could give Beijing a decisive edge in any future conflict with Taiwan. An overwhelming offensive advantage could intensify the existing cross-Strait arms race, reduce Beijing's willingness to compromise on cross-Strait issues, increase the chances that force could be used short of an outright Taiwan declaration of *de jure* independence, and naturally prompt Taiwan to shift toward a tactically offensive doctrine. At the extreme, an overwhelming PLA offensive advantage could force Taiwan to pursue a punitive deterrent option.

In this chapter, the author will first address drivers that are influencing the PLA force planners who view space assets and theater missiles as integral to 21st century operations. He next outlines Chinese efforts to field a space-based reconnaissance architecture that would support theater missile targeting; then he reviews research and development aimed at fielding a large arsenal of ballistic and land attack cruise missiles. Next, the author details operational concepts associated with a PLA theater missile campaign, to include organizational issues, information denial, and the Second Artillery's phased approach to theater warfighting. He concludes with a discussion of the operational and political implications of an offense-dominated force structure, as well as potential countermeasures.

DRIVERS

A number of drivers are propelling Beijing toward reliance on theater missiles and supporting space assets. These include: (1) lessons from the Gulf War and subsequent U.S. and Russian literature on the revolution in military affairs (RMA); (2) a doctrinal shift toward offensive preemption, surprise, and deep strikes against strategic and operational targets; (3) use of Taiwan as a preeminent force planning scenario; and (4) prevention of foreign intervention in a Taiwan scenario through an area denial strategy.

Lessons from the Gulf War. China's interest in deep attack was sparked in large part by lessons from the Gulf War and subsequent U.S. and Russian literature on the revolution in military affairs (RMA). The U.S. performance in the Gulf War demonstrated to the Central Military Commission (CMC) the preeminence of the offensive, especially airpower and long-range precision strike. In a December 1995 meeting, the CMC concluded that "ground fighting can only enhance the results of battle." Lessons from the Gulf War have been reinforced by calls to meet its challenges of 21st century warfare through selected exploitation of the RMA.[3] Chinese commentators note areas for exploitation include precision strike, strategic maneuver, and space combat.[4]

Emerging Operational Concepts. The Gulf War and the RMA have sparked a fundamental reassessment of the PLA's approach to warfare. Operational concepts articulated in a wide range of PLA publications serve as an important driver propelling the PLA toward theater missiles and supporting space assets. Key to future conflicts around the PRC's periphery will be achieving a political resolution through rapid establishment of information dominance (*zhixinxiquan*) and air superiority (*zhikongquan*) in the opening phases of a conflict.[5] The concept of "rapid war, rapid resolution" (*suzhan, sujue*) requires a series of crippling strikes directed against vital points (*dianxue*) of the enemy's defense infrastructure. These critical nodes include civilian and military command and control facilities; intelligence, surveillance, and reconnaissance nodes; and important airfields and air defense sites. This does not require annihilation of the enemy or occupation of his territory, only a paralyzing "mortal blow" (*zhiming daji*), "winning victory with one strike" (*yizhan, ersheng*).[6] From the Chinese perspective, "gaining the initiative by striking first" (*xianfa zhiren*), is one of most effective means of offsetting the technological and logistical advantages that a more advanced military power would bring to the fight. The emerging doctrine requires a high degree of secrecy, mobility, an accurate concentration of firepower, and surprise.[7]

Use of Taiwan as a Primary Force Planning Scenario. Since the collapse of the Soviet Union, Taiwan appears to have become a primary testing grounds for the PLA's emerging operational concepts. Military force planners around the world generally rely on illustrative planning scenarios to guide the development of doctrine, research and development (R&D), and acquisition. Until the early- to mid-1990s, China did not appear to be fostering an ability to take Taiwan by force. Nor did the PRC deploy more than a symbolic land, sea, or air force presence within 300 miles of Taiwan. Now, however, PLA modernization—and theater missile development in particular—is motivated in large part by the desire to use decisive military force as a means to deter Taiwan independence sentiment and strengthen the PRC's hand in a reestablished cross-Strait dialogue. The focus on Taiwan may reflect a view within the PLA that force may eventually have to be used.

With Taiwan as the primary driver, the PLA has three general operational requirements. First is the capacity to bring Taiwan to its knees quickly through paralysis of Taipei's ability to conduct military operations. Critical to this effort is establishment of information dominance by neutralizing Taiwan's intelligence, surveillance, and reconnaissance assets and paralyzing its command and control network. Information dominance enhances the conditions necessary to control the airspace over Taiwan. Theater ballistic and land attack cruise missiles, used in parallel with electronic warfare, special operations, and offensive counterair operations, can play a crucial role in the rapid establishment of information and air superiority. Control of the information environment and the skies above the Taiwan Strait—if not enough to force a resolution in itself—could create the conditions necessary for dominance of seas and facilitate an amphibious invasion, if necessary.[8] The PLA must also hedge against strikes against its own critical assets and facilities.

Prevention of Foreign Intervention. At the same time, the PLA must deny foreign forces the ability to intervene either through a quick resolution of the conflict or through complicating their ability to enter the area of operations. Since the U.S. deployment of two aircraft carrier battle groups off the coast of Taiwan in March 1996, PLA planners probably assume the United States would intervene in a future Taiwan scenario. PLA writings indicate Beijing is pursuing the kinds of capabilities intended to deter or prevent intervention by outside powers such as the United States. The PLA has carefully studied U.S. military weaknesses and has identified vulnerabilities in U.S. force projection, including reliance on space systems, weaknesses in aircraft carrier battle groups, and expeditionary air forces.[9]

The most fundamental requirement for denying the United States the ability to intervene in a Taiwan conflict would be an expanded capacity for intelligence, surveillance, and reconnaissance. Monitoring U.S.

deployments could enable PLA targeting of critical nodes in the Western Pacific in order to complicate or delay U.S. intervention in a Taiwan scenario. Successful use of overwhelming force through preemptive strikes to quickly resolve the Taiwan issue could preclude U.S. intervention by presenting Washington and the international community with a *fait accompli*.

SPACE SUPPORT FOR THEATER MISSILE OPERATIONS

Under CMC guidance, China's space and missile industry intends to field a constellation of reconnaissance systems that could support the PLA with near-real-time intelligence early in the 21st century. PLA observers view exploitation of space assets as crucial for 21st century warfare. Theater operations must be supported by surveillance architecture for strategic intelligence, targeting, and battle damage assessment (BDA). Effective theater missile operations need to see deep. Before any targets can be struck, they must be identified as targets, precisely located, and defenses accurately assessed so that they can be hit without prohibitive losses. This requires information from a variety of space-based, airborne, and ground-based sensors. Through its existing air- and ground-based reconnaissance network, the PLA currently has the ability to monitor activities within line of sight of its borders—approximately 200 nautical miles.[10] However, to expand its battlespace awareness, the PLA must develop the means to monitor activities in the Western Pacific, South China Sea, and Indian Ocean. Space assets could enable the monitoring of naval activities in the Pacific and Indian Oceans and the South China Sea and track U.S. Air Force expeditionary air force deployments into the region. Space-based reconnaissance systems also provide the images necessary for mission planning functions, such as navigation and terminal guidance for land attack cruise missiles.

China Aerospace S&T Corporation (CASC) is developing at least four space-based systems that would expand PLA battlespace awareness and support strike operations further from Chinese shores.[11] Space operations are the responsibility of the PLA General Armaments Department (GAD) China Launch and Tracking Control General (CLTC).[12] While only a small percentage of space-derived intelligence, surveillance, and reconnaissance (ISR) assets will be near real time, the number and diversity of sensors could provide frequent revisit times and complementary data on significant military targets on Taiwan and in the Western Pacific.[13]

By the 2005-2010 timeframe, China's space-based surveillance architecture could have at least four components: 1) synthetic aperture radar (SAR) satellites for all weather, day/night monitoring of military activities; 2) electronic reconnaissance satellites to detect electronic emissions in the Western Pacific; 3) mid-high resolution electro-optical satellites for early warning, targeting, and mission planning; and 4) a new generation of high resolution recoverable satellites for intelligence and analysis. According to Chinese sources, SAR and electronic reconnaissance satellites would serve as important components of an ocean monitoring (*haiyang jianshi*) network for detecting and tracking naval activities, to include carrier battle groups and submarines. Development of space-based surveillance architecture has in large part been funded under the special 863-program budget.[14]

Radar Imaging Satellites. Work on an indigenous synthetic radar (SAR) satellite (*hecheng kongjing leida weixing*) began in the 1980s. Under the 863 Program, China's space industry and oceanographic research organizations began preliminary research on a SAR satellite in 1987. The program moved into the applied R&D phase in 1991. After successful fielding of an airborne SAR system,[15] China's State Science and Technology Commission (SSTC) and the PLA's Commission on Science,

Technology, and Industry for National Defense (COSTIND) approved the finalized design and associated high speed data transmission system in May 1995.

Production of the first generation SAR satellite is included in the 9th Five-Year Plan (1996-2000). China's first radar imaging satellite, designated the Haiyang-1 (HY-1), is slated for launch in 2001. The HY-1 will be based on a small satellite bus that will serve as a common bus for a range of future satellite constellations, to include an integrated SAR/Electro-optical (EO) small satellite constellation. The HY-1 and major subsystems passed a design finalization review recently and a test model is supposed to be delivered by end of this year. Preliminary research has already begun on a more sophisticated second-generation SAR satellite system.[16]

While SAR satellites have civilian applications, Chinese journals note their principal purpose is to support national defense. The PLA and other parts of the state apparatus view radar satellite imagery as a critical modernization program. Unlike electro-optical systems, GSD Second Department advocates note that space-based SAR systems can see through clouds, rain, and fog in order to detect and track ships and submarines in shallow waters.[17]

China has arranged to receive downlinked radar satellite imagery to help establish a foundation for radar satellite imagery exploitation. The PRC has entered contractual agreements to obtain commercial radar satellite data from a number of foreign vendors. China began receiving SAR data from ERS-1 and JERS-1 satellites in 1994 and from Canada's RADARSAT in 1997. Included in the arrangement was training of imagery analysts.[18]

Electronic Reconnaissance Satellites. Electronic reconnaissance satellites (*dianzi zhencha weixing*) appear to be the second component of an ocean-monitoring network. Strong indications exist that China has resurrected an electronic reconnaissance satellite program that has been dormant for over 20 years. The PLA experimented with electronic reconnaissance satellites, euphemistically called "technical experimental satellites" (*jishu shiyan weixing*), in the mid-1970s under the Shanghai Bureau of Astronautics' 701 Program. Technical writings indicate the Shanghai Academy of Spaceflight Technology (SAST), the successor of the Shanghai Bureau of Astronautics, has resurrected the program and intends to field a space-based electronic reconnaissance system. At least one SAST design under evaluation is a constellation of small electronic reconnaissance satellites that can ensure precise location data and survivability.[19]

Electro-optical Reconnaissance Systems. In addition to its ocean reconnaissance systems, China's remote sensing community is working feverishly to deploy space-based electro-optical (EO) remote sensing platforms. CASC and China's electronics industries have made notable progress in charged couple devices (CCDs), a technology which is essential to the development of real-time EO imaging systems.[20] Fielding of EO satellites would enable Beijing to beam images back to ground stations directly from space.

The Ziyuan-1 (ZY-1), a joint venture between the PRC and Brazil, will serve as China's first EO reconnaissance satellite. Launched in October 1999, the ZY-1 has a 2-year lifespan and incorporates a data transmission system to beam images back to earth. The ZY-1, operating at an altitude of 778 kilometers, is expected to have only a 20-meter resolution but will add to China's experience base in EO imaging systems.[21]

Building on the experience gained from the HY-1 and ZY-1 programs, CASC spokesmen have announced their intention to field a tactical small satellite imaging constellation and associated mobile ground receiving stations. The tactical imaging system, slated for launch in next two years, will consist of four EO and two SAR satellites. The EO component likely will use the same bus as the HY-1 and is designed to have a five-meter resolution when operating in a 700-kilometer orbit.[22]

Small satellite constellations are an important aspect of China's operational concept for space warfare. Clearly recognizing their military implications, Chinese defense officials advocate small satellite development in order to reduce vulnerability of fixed launch sites. Chinese engineers are examining the utility of using mobile, solid fueled launch vehicles, such as a modified DF-21 or DF-31.[23] Reduced size and complexity allows for faster R&D and manufacturing time, and production in significant numbers. In a contingency situation, tactical imagery satellites can be launched on demand, with mobile launch platforms increasing survivability. Multiple small satellites can be stacked on a single launch vehicle. Furthermore, enemy attacks on small satellite constellations will encounter greater targeting difficulties and be costly. Destruction of one satellite will have minimal effect on the overall functioning of the architecture.[24]

The FSW-3. China has launched more than a dozen film-based recoverable satellites (*fanhuishi weixing, or FSW*) since 1975. These systems stayed in orbit for up to 16 days and were used to obtain imagery of Taiwan and nations around China's periphery, determine coordinates of facilities that were potential targets of Chinese missiles, and map Chinese territory. China's most recent generation of reconnaissance satellite, the Fanhuishi-2 (FSW-2), displayed an ability to maneuver in orbit. Reconnaissance satellites generally have been launched from Gansu's Jiuquan Space Launch Center.[25] China's next generation recoverable satellite—the FSW-3—is expected to have a resolution of one meter. This satellite could be launched as early as later this year.[26]

Ground Processing. China's ground processing capacity is rapidly progressing. Chinese engineers are working to improve ultra high-speed data processing, storage, and transmission systems, as well as computer, data compression, and networking technology to be able to handle real time, high-resolution imagery from multiple satellites. Essential for the efficient transmission and processing of satellite-derived imagery is data compression technology, which CAST is attempting to master.[27] The PLA has fielded a real time image storage system as well as an imagery dissemination system that is linked to China's national integrated telecommunications network. The system will allow subscribers to search and rapidly download images.[28] In 1996, the PLA installed a digitized high-resolution imagery processing system, the BGC-161.[29]

China is receiving foreign assistance. In 1992, Italy's Telespazio signed an agreement worth U.S. $8 million to provide Olivetti image processing computers and software. Telespazio assigned technicians to train Chinese photo interpreters for up to three years. China's procurement of foreign sources of imagery also includes options for training.[30]

Despite significant investment in reconnaissance systems, China may still have a limited near-real-time targeting capability. Reconnaissance satellites must be within line of sight of a ground station to download their imagery data. Targets on Taiwan could be imaged and immediately beamed back to a ground station on the mainland. However, satellites imaging targets further out from China's borders in the Western Pacific likely would need to store their images and wait until they return to within line of sight of the Chinese mainland.

Future deployments of a sea-based imagery receiving station, a data relay satellite (DRS), or establishment of ground stations abroad would enhance China's extended range near-real-time targeting capability.[31] A Chinese DRS architecture under development is expected to include at least two geostationary satellites that could provide 85 percent coverage of the earth and support 5-10 satellites at the same time.[32]

R&D, production, and deployment of satellite systems are expensive. However, much of the R&D budget for China's space program comes from the State Council science and technology budget, not from PLA coffers.[33] With a price tag of between 5-12 million U.S. dollars per satellite, the cost of satellite development in China is significantly less than in the United States or Western Europe.[34] In addition to funding from the 863 Program, R&D of space systems is subsidized by revenues from commercial space launches and the sale of satellite systems abroad. International cooperative efforts with Russia, Ukraine, Belarus, France, Germany, Italy, and Brazil cut costs even more.[35]

THEATER MISSILE DEVELOPMENTS

A space-based reconnaissance system will be a key element of the PLA's emerging theater missile strike force. In March 1996, the Central Military Commission reportedly convened an enlarged meeting and developed a plan to develop seven weapons on a priority basis. Four of those weapons are directly related to building a deep strike capability. At least one of the objectives was fielding China's first generation land attack cruise missile by 2000.[36]

Dependence on theater missiles reflects a failure of China's aviation industry to provide the types of aircraft that normally would carry out this mission.[37] Although they can carry only one-sixth the payload of an air-to-ground strike fighter, ballistic missiles have a strong psychological deterrent effect, and are increasingly accurate, mobile, and stealthy. Advocates argue that ballistic and land attack cruise missiles are relatively cheap while aviation technology is increasingly sophisticated and expensive. Air mobilization is time consuming and relatively easy to detect. Air strikes against targets in denied areas require a measure of air superiority. Theater ballistic missiles, however, can be rapidly mobilized with a high degree of secrecy. They are much harder to counter due to their fast reentry speeds and short flight times.[38]

Theater Ballistic Missiles.

Ballistic missiles are emerging as the backbone of conventional PLA theater operations. Drawing profound lessons from the Gulf War, the PLA views conventional ballistic missiles as a crucial aspect of China's emerging deep attack (*zongshen daji*) strategy.[39] The China Aerospace S&T Corporation (CASC) appears to be producing a substantial number of conventional theater ballistic missiles with ranges stretching from 300 to 2000 kilometers. In fact, a 1998 Department of Defense report asserted that China's space and missile industry probably will have the capacity to produce as many as 1000 ballistic missiles in the next decade.[40] At an estimated cost of U.S. $500,000 or less per missile, CASC would be able to produce up to 1000 ballistic missiles at a total cost of $500 million.[41] CASC's key producers of ballistic missiles—China Academy of Launch Technology and the 066 Base in Hubei province—are leveraging foreign technology in order to achieve tremendous advances in accuracy. At the same time, they are diversifying the payloads of their ballistic missile to increase their lethality. CASC and the PLA are also examining a wide range of countermeasures to ensure their theater ballistic missile force remains viable as active theater missile defenses are introduced into the Asia-Pacific region. The PRC is concentrating on three conventional theater ballistic missile systems: 1) the DF-15 short-range ballistic missile (SRBM); 2) the DF-11 SRBM; and 3) the DF-21 medium-range ballistic missile (MRBM).[42]

DF-15 Short-Range Ballistic Missile System. The DF-15 (CSS-6) is a solid-fueled, 600 kilometer SRBM manufactured by the China Academy of Launch Technology (CALT). The DF-15's payload reportedly has an attitude control mechanism that permits steering corrections from separation to impact.[43] The detachable warhead offers a much smaller target than a SCUD, and its potential maneuverability would complicate missile defense radar tracking, computations, and interception. With a unitary, high explosive warhead, the DF-15 could create a crater as large as 30-50 meters in diameter.[44] Assuming a nominal trajectory at a range of 500 kilometers, the DF-15 would reach an altitude of about 120 kilometers, achieve a reentry speed of about two kilometers per second, and have a flight time of only six or seven minutes.[45] Some reporting indicates the DF-15 currently has a 100-meter circular error of probability (CEP).[46] To diversify its theater ballistic missile inventory, a 1200-kilometer range version of the DF-15 is reportedly under development.[47]

DF-11 Short Range Ballistic Missile System. The DF-11—better known by its export designator, the M-11 (CSS-7) — also is a solid propellant, road-mobile SRBM with an estimated range of 300 km. This missile, however, has not yet entered the PLA's inventory. An improved, longer-range version of the DF-11 may be under development.[48] The main advantage of the DF-11 over the DF-15 is its ability to carry a larger payload. Some sources credit the 300-kilometer version with an 800-kilogram warhead and a 150-meter CEP.[49] The DF-11 is manufactured by the CASC's 066 Base, also known as the Sanjiang Space Corporation, based in Hubei province. The DF-11's 300-kilometer range presents challenges for active missile defenses due to its brief flight time of three minutes. Because its flight would remain within the atmosphere, upper tier systems would be unable to engage the 300-kilometer DF-11.[50] Details on the longer-range version are unavailable.

DF-21 Medium Range Ballistic Missile System. One other missile system which could be brought to bear against Taiwan is the solid-fueled 2000 kilometer DF-21 (CSS-5), equipped with a 600 kilogram warhead. Research and development on the DF-21 began in 1967, had its first successful test in 1985, and deployed into an experimental regiment as early as 1991. With a circular error probable

(CEP) of 700 meters, the DF-21 is currently equipped for nuclear missions only. However, there are indications that since 1995 a terminal guidance system has been under development for the DF-21 that could permit highly accurate conventional strikes.[51] The DF-21 reentry speed is fast enough to preclude successful intercepts by lower-tier missile defense systems. Because of its warhead size and the inability of lower tier missile defense systems to engage longer range MRBMs, incorporation of a terminal guidance system could have significant military implications.

Technical Trends.

Several efforts are underway to increase the accuracy and lethality of China's theater ballistic missiles. These include: 1) terminal guidance; 2) a diverse mix of conventional payloads; and 3) missile defense countermeasures.

Terminal Guidance. The most significant development in China's theater ballistic missile program is the development of terminal guidance systems that, according to Chinese writings, meet a CEP requirement of 25-40 meters.[52] CASC engineers point to three options in ballistic missile terminal guidance. First, terrain matching terminal guidance (*dixing pipei mozhidao*) makes use of digitized stored images (electro-optical or radar) and matches them against the images acquired in the seeker. CALT began preliminary research on terrain matching terminal guidance as early as 1977.[53] Radar matching was used on the U.S. Pershing-II and optical matching is currently used on a Russian variant of the SCUD-B. Chinese engineers note that critical technologies for terrain matching terminal guidance include large scale and very large-scale integrated circuits (LSIC/VLSIC).[54]

A second means for terminal guidance is a millimeter wave seeker (*maomibo xun*). CALT engineers have carried out a number of feasibility studies on terminal guidance technologies, to include millimeter wave and infrared.[55] Millimeter wave seekers are compact, lightweight, have high spatial resolution, a robust antijamming capability, and are all weather. Critical technologies include LSICs, microcomputers, and digital information management systems for target discrimination and tracking. Chinese engineers, however, note that millimeter wave (MMW) seekers are relatively expensive.[56]

A final option for terminal guidance is exploitation of the global positional system (GPS). GPS-assisted guidance systems usually include a GPS receiver, a ring laser gyro (*huanxing jiguang tuoluo*), and microcomputer. There are indications China has already mastered use of GPS for mid-course corrections. At least two tests of an on-board GPS trajectory reference system had been conducted as of 1995.[57] Use of GPS for terminal guidance requires frequent and highly precise updates from navigation satellites. Potentially in support of this effort, China is installing a differential GPS network (*chafen quanqiu dingwei xitong*) along its eastern seaboard that could enhance the accuracy of the PLA's SRBM force.[58]

As a final note, CASC institutes have close relations—some officially sanctioned and some not—with counterparts in the former Soviet Union. Chinese engineers have approached Russian institutes for ballistic missile guidance and control technology and have hired a number of Russian scientists as technical advisors.[59]

Conventional Payloads. There is evidence that China intends to design up to six different payloads for its theater ballistic missiles. Today, the PLA's conventional theater ballistic missiles are armed only with unitary blast fragmentation warheads. To diversify its missile payloads, CASC writings indicate prioritization of three categories of specialized warheads for use against air defense sites, radar, airfields, semi-hardened C4I centers, and ports: 1) submunition (*zimudan*) payloads; 2) electromagnetic pulse warheads (*dianci chongzhong dantou*); 3) penetrating warheads (*zuandi dantou*); and 4) fuel-air explosive warheads (*youqi* or *leibao dantou*).[60]

Submunitions. A submunition warhead contains a number of small devices or "bomblets" designed for specialized roles. These warheads usually detonate at a preset altitude of several hundred meters so as to spread the munitions out to an optimal pattern size. Submunition warheads are far more efficient against targets susceptible to blast and fragmentation than unitary warheads of the same weight. As of 1996, CALT was testing a guided submunition (*jiandan zimudan*) package for blast and fragmentation effects; and penetrating submunitions (*qinche zimudan*) for cratering runways. More advanced packages under development include terminally guided sensor fused submunition warheads.[61] There are also indications of CBU-78 GATOR-like mine laying submunition development.[62]

Electromagnetic Pulse Warhead (EMP). PLA writings indicate that fielding of an anti-radiation EMP warhead is a high priority. An anti-radiation warhead, specifically a high-powered microwave (HPM) device, is viewed as a "natural enemy" (*tiandi*) of more technologically advanced militaries and an "electronic trump card" (*dianzi shashou*).[63] Due to challenges related to weaponizing a device with enough power, HPM warheads would initially only be effective against radiating targets within the immediate area of impact—radar and communications centers would be the prime candidates. As the technology progresses, HPM warheads could achieve wider effects.[64] The developers of the DF-11—known as the 066 Base—have demonstrated the most interest in HPM warheads.[65]

Penetration Warheads. Chinese engineers note that an increase in CEP to better than 50 meters would permit the use of penetration warheads (*zuandi dantou*) that would dig deep into semi-hardened facilities such as command and control centers, intelligence collection facilities, and weapons storage facilities which are housed in concrete bunkers. CALT warhead engineers have tested a range of high strength steels and other material and structural technologies that would dig into critical hardened facilities.[66]

Fuel-Air Explosive Warhead. There are also indications that the DF-15 may carry a fuel-air explosive (FAE) warhead. FAE warheads offer greater explosive power at a weight approximately 50 percent less than conventional explosives. Pound for pound, FAE weapons are three to five times as destructive as high explosive warheads. For example, a 500kg FAE warhead would destroy most aircraft and injure all personnel within 250 meters of the impact point. Chinese designers have studied the use of FAE warheads since the 1970s and tested the effectiveness of an FAE as early as 1976 by detonating an U.S. device that had been captured by the Vietnamese and transferred to Beijing.[67]

Missile Defense Countermeasures.

The PLA places a premium on ensuring its ballistic missile force would be able to penetrate future active theater missile defenses. PLA and defense industry analysts are examining a range of more sophisticated TMD countermeasures that could reduce the effectiveness of active missile defense systems.

Saturation. The simplest means of overcoming missile defense architecture is through saturating it with a large number of missiles. Given enough ballistic and land attack cruise missiles, any system can be saturated by overwhelming a missile defense systems' area of coverage. Saturation generally requires a large number of missiles, timed to arrive together in order to bunch effectively for ground defense saturation. PLA General Armament Department engineers have evaluated PATRIOT saturation rates and are confident that their theater ballistic missiles can reach their targets.[68]

Maneuvering Re-Entry Vehicles. More sophisticated countermeasures could reduce the effectiveness of active missile defenses short of the brute force saturation approach. For example, the CASC is developing the capabilities that would permit conventional ballistic missile re-entry vehicles to maneuver in their terminal phase. Missile designers believe maneuvering not only is essential for terminal guidance packages but is also a means to complicate ballistic missile defenses. Through modeling and simulation, CASC has determined that maneuvering is a viable means to reduce land-based lower tier TBMD probability of kill rates.[69] In support of this research effort, China allegedly acquired PATRIOT technology to calibrate an auxiliary propulsion system on the DF-15 re-entry vehicle to enable the payload to outmaneuver a PATRIOT system as it re-enters the atmosphere.[70] Missile designers have also demonstrated a special interest in the speed control maneuver (*sudu kongzhi jidong*) used in the 1800-kilometer Pershing-II.[71]

Shaping, Stealth, Decoys, and On-Board Jammers. CALT warhead designers have already lowered the DF-15 and DF-11 signature through shaping of the warhead. A warhead designer can lower the signature of a reentry vehicle further by reducing the infrared signal or by incorporating stealth design technologies to reduce the radar cross-section. Use of balloons can also mask the shape of reentry vehicles and chaff (*jinshu botiao*) can be released with the reentry vehicle in an attempt to hide the target behind a cloud of radar reflecting metal strips. Chinese engineers have tested chaff packages. Other measures under investigation include electronic and infrared countermeasures on board reentry vehicles, as well as carrying out hard kills against enemy TMD radar through the use of anti-radiation missiles.[72] CASC missile engineers have tested active jammers that can broadcast a signal designed to interfere with a radar's ability to detect the target object or corrupt the signal in such a way as to cause the radar to receive a false echo.[73] National University of Defense Technology analysts have examined electronic countermeasure packages on board theater ballistic missiles as a means to counter millimeter wave and infrared seekers on missile defense interceptors.[74] In 1995 and 1996, the Chinese allegedly tested DF-21 endo-atmospheric decoys.[75]

Laser Cladding. Looking ahead to the potential deployment of boost phase intercept systems such as the airborne laser (ABL), Chinese engineers are developing a coating for ballistic missiles that could complicate use of missile defense high power lasers. Using their own indigenously developed high powered lasers, Chinese institutes have tested various coating materials to protect the outer shell of ballistic missiles, a process known as laser cladding (*jiguang rongfu*).[76] Laser cladding, together with the spinning of theater ballistic missiles, may not make ballistic missiles immune to boost phase missile defense systems but could increase required lasing time, thus reducing the number of laser shots available per ABL mission.

Multi-Axis Attacks. The Second Artillery and CASC have conducted modeling exercises and simulations to test China's ability to break though the wide range of projected TMD deployments. Modeling has focused on large raid sizes, using combinations of surface-to-surface, air-to-surface, and sea-to-surface theater missile systems. After computer simulations and modeling exercises, CASC is confident that its theater ballistic missiles can neutralize opposing land based lower tier systems.[77]

Depressed Trajectories. Chinese missile analysts view depressed trajectories (*yadi guidao*) as another option to counter space-based and exo-atmospheric upper tier missile defense systems. A 1200-kilometer range ballistic missile on a nominal trajectory will normally reach an altitude of 400 kilometers, rendering the missile vulnerable to upper tier missile defenses for a substantial portion of the flight. However, launching a missile at a depressed trajectory could allow the missile to achieve only a 100-kilometer altitude that limits the ability of exo-atmospheric upper systems to engage the missile. Testing and modeling has been done on the DF-3 (CSS-2), which normally has a range of 2780 kilometers, with a maximum altitude of 550km. With depressed trajectory, the DF-3 travels 1550 kilometers at 100 kilometers altitude.[78]

Land Attack Cruise Missiles.

To augment its theater ballistic missile arsenal, China is creating a new generation of cruise missiles able to penetrate defenses and strike critical targets with precision and increased firepower. Fielding of land attack cruise missiles (LACMs) may prompt expansion of missions of the PLA's Second Artillery and Navy. Increasing availability of cheap navigation and guidance systems and digital mapping technology have increased the incentives and reduced the time required to field a LACM.

Cheaper and more accurate than ballistic missiles, LACMs appear to have a relatively high development priority. The size and flight profile of ground-, air-, and sea-launched cruise missiles can stress the capabilities of even the most modern air defense systems. Chinese research and development of LACMs is being aided by an aggressive effort to acquire foreign cruise missile technology and subsystems, particularly from Russia. The first LACM to enter production probably would be air-launched and could be operational early in the 21st century.[79]

The heart of China's LACM missile development lies within CASC's Third Academy, headquartered just southwest of Beijing. Over 14,500 technicians and workers ply their trade in ten research institutes and two major factories. The following discussion of China LACM development focuses on: 1) the underlying rationale for LACM development; 2) specific LACM systems that may come on line within the next 5 years; 3) the mission planning process; and 4) general technical trends influencing China's LACM development.[80]

Why Land Attack Cruise Missiles? Land attack cruise missiles have a number of advantages over ballistic missiles or manned aircraft. China's first generation LACM is likely to be up to twice as accurate as their theater ballistic missiles. Successful exploitation of GPS, indigenous and/or foreign procured remote sensing data, and digital mapping technology could permit the fielding of an LACM with a CEP of 16 meters or better. LACMs are cheaper to produce— generally thought to be one-third the cost of ballistic missiles. For example, if one assumes an SRBM unit cost of $500,000, then the unit cost of an LACM could be as little as $175,000. Chinese defense industrial observers note that developing an arsenal of cruise missiles could have a 9:1 ratio over the cost of defending against them. As the president of the Third Academy has pointed out, the cost of producing cruise missiles is 20-30 percent less in China than it is in other countries.[81]

Cruise missiles offer other appealing features as well. Use of GPS allows launchers to forego pre-surveyed launch sites, permitting the missile to disperse to a greater range of launch sites. Ground-launched LACMs can be quite survivable. With a low take-off weight, they tend to be more easily transportable than theater ballistic missiles. The infrared launch signature would be less than that of a ballistic missile, decreasing warning time and increasing survivability. Unlike ballistic missiles, land attack cruise missiles could be loaded onto ships and fired at land targets that may have not been anticipated.[82]

Land attack cruise missiles pose serious challenges for air defenses. Due to the earth's curvature, ground-based radar can detect a low flying cruise missile only about 32 kilometers away. In comparison, an aircraft flying at 10,000 feet can be detected when it is about 240 kilometers away. Newer missiles are incorporating stealth features that would make them even less visible to radar and infrared detectors.

PLA LACM Programs. In accordance with standard Third Academy R&D practices, China's future family of land attack cruise missiles likely will be based on airframes which have already been fielded.[83] Chinese and Western sources indicate at least three families of cruise missiles may be under evaluation for a land attack mission: 1) the Silkworm; 2) the multipurpose Yingji-8 missile; and 3) an anti-radiation missile that Western sources have designated as the YJ-9.[84]

The Land Attack Silkworm. China's first LACM is expected to be a Silkworm derivative. This system, designated the XY-41 as early as 1989, will be smaller, more mobile, and more accurate than ballistic missiles such as the DF-15, but carry the same size warhead (500kg). The XY-41 is a variant of the HY-4 anti-ship cruise missile.[85] The Silkworm-derivative could be air- or ground-launched and reportedly will have a range of 300-400 kilometers, indicating an upgrade to the HY-4's turbojet engine.[86] Some Western reporting asserts that CASC is getting foreign assistance in development of an integrated INS/GPS system and in warhead technology.[87]

Ground-launched LACMs would be subordinated to the Second Artillery. Based on existing organizational structures within the Second Artillery and in coastal Silkworm units, a ground launched LACM brigade likely would be divided in four battalions, with each battalion having four company-sized fire units with one launcher per fire unit. A first generation LACM brigade could adopt an organization structure similar to today's typical HY-4 fire unit—four towed launchers, a firing command vehicle, a truck mounted microwave relay, and auxiliary power truck. The HY-4 is launched via a solid rocket booster before a turbojet engine takes over for the duration of the flight. The missile cruises at about Mach 0.8 and maintains an altitude between 70-200 meters.[88]

The YJ-8. A second system reported by Western sources as being adapted for land attack use is the smaller Yingji-8

(YJ-8). The Exocet-like YJ-8 series adopts a solid propellant that decreases the size and weight of the system to enable more diverse set of launching modes. With the solid motor, however, the YJ-8's range is limited to 42 kilometers or less. A YJ-8 follow-on, designated the YJ-82, utilizes the same basic airframe but uses a small turbojet engine instead of a solid motor that extends the airframe's range to 120 kilometers. The alleged land attack version of the YJ-8—reportedly the turbojet variant—would incorporate integrated GPS and TERCOM guidance. Western sources indicate the Third Academy may be extending its range to at least 300 kilometers and claim that GPS-aided navigation—augmented by terrain contour matching—could result in cruise missiles like the YJ-8 to achieve an accuracy of up to 10 meters. The YJ-8's smaller warhead (165kg), however, would limit the missile's utility.[89]

A Chinese Anti-Radiation Missile. The PLA and China's space and missile community have also devoted a significant degree of attention to development of cruise missiles with passive seekers to counter enemy radar systems. Like ballistic and other cruise missiles, anti-radiation missiles are considered as a *shashoujian* for priority development. Western sources have designated China's new family of anti-radiation missiles as the Yingji-9 (YJ-9), a system influenced by Russia's Kh-31P and/or Israel's STAR-1 ARM systems. There are persistent rumors of PLA procurement or joint production arrangement on the Kh-31P, which Chinese engineers note was specifically developed to counter the U.S. AWACS, PATRIOT MPQ-53 radar, and AEGIS SPY-1D phased array radar. China's defense industrial complex, specifically the Third Academy with support from the Harbin Institute of Technology, is aggressively pursuing deployment of a long-range anti-radiation missile.[90] Some Western sources allege an extended range version of the YJ-9 may have a range of 400 kilometers.[91] Chinese research indicates China's first generation anti-radiation missile will be air-launched.[92]

Mission Planning. Mission planning exploits navigation aids and flight management computers to permit LACMs to fly along precise, pre-programmed routes. This generally involves use of a land attack navigation system, including exploitation of U.S. NAVSTAR GPS and Russia's GLONASS, a radar altimeter, an inertial measurement unit, and a sophisticated flight management computer. Because mission planning requires a knowledge of the shape of the terrain and obstructions found along the cruise missile's intended flight path, satellite imagery and geographic information systems (GIS) play an important supporting role.[93]

Much of this technology is available commercially off-the-shelf.[94] Commercial imagery is adequate to plan routes with relative positional accuracy on the order of tens of meters. China's indigenous remote sensing program and future commercial sources will provide even more precise data. There are a number of commercially available mission planning software programs which can manipulate sources of imagery for robust mission planning for military purposes.

There are two major mission planning processes: en route and terminal. Both heavily rely on intelligence. For the enroute planning process, General Staff Department (GSD) intelligence and cartography/mapping offices probably would identify enemy air defenses to avoid and generate en route terrain data. Routes would be validated and waypoints en route to the target loaded into the LACM computer.[95] Because of the requirement for large databases and computer operations, the mission planning process likely would be centralized in Beijing and then transmitted to the theater operations command center.

An important en route mission planning technology under development is terrain contour matching (TERCOM). A prerequisite for TERCOM (*dixing pipei zhidao*) is generation of electronic maps from high-resolution satellite images. TERCOM uses a radar altimeter to measure terrain features along its flight path and correlate these measurements with internally stored digital maps. The Third Academy has been conducting preliminary research into TERCOM since at least 1988.[96] There is some indication China is examining integrating combined GPS/GLONASS receivers on board their missiles as well.[97]

For terminal planning, the most advanced PLA system under development is the digital scene matching area correlation (DSMAC) system. DSMAC updates the position of the missile by matching a stored image to a series of images sensed in flight. The planning required is substantial and complex. A PLA targeteer likely would obtain a photograph of the targeted area and generate DSMAC scenes for programming into the cruise missile's flight computer. The DSMAC images are prepared from satellite photographs of the target. Higher resolution images allow for more accurate updates and a better CEP. PLA GSD intelligence analysts would identify targets of interest and then either pull the image from the library or task China's remote sensing community to procure the image. PLA GSD targeteers would then select aimpoints to exploit the most vulnerable aspect of a command and control facility or airfield. The photograph would then be transformed into a digital image and loaded onto the LACM computer. Third Academy engineers believe en route and terminal mission planning systems can assure a CEP of 16 meters or less.[98]

The mission planning process can take up to several hours. The PLA's deep attack strategy, however, does not necessarily require a rapid mission planning process since most targets on Taiwan would be strategic and static in nature. If indigenous near-real-time space remote sensing systems were available, GSD would task the satellite operators for the imagery. The GSD could also order quick-turnaround imagery from a foreign provider, such as Russia, Israel, or France.

Deployment of a LACM capability likely will spark a shift in organizational responsibilities. As previously mentioned, the Second Artillery appears to be a primary procurement agent for ground launched LACMs.[99] An air launched Silkworm LACM will likely use PLA Air Force subordinated B-6s. If the YJ-8 were fielded as a LACM, tactical fighter-bombers such as the FB-7 would serve as the primary platform. There are indications the PLA Navy may be seeking to expand its mission to include conventional missile strikes against land-based targets from the sea.[100]

Technical Trends.

China's Third Academy intends to upgrade its cruise missile production capacity, extend the range and speed of its cruise missiles, diversify its choices of terminal guidance systems, and lower their radar cross-section.

Production Trends. The Third Academy is upgrading its ability to design and manufacture highly complex cruise missiles. They are integrating the use of virtual reality (*xuni xianshi*) in cruise missile development, and are using increasingly sophisticated supercomputers to design the missiles. Third Academy manufacturing centers have imported some of the world's most advanced engineering workstations, and three, four, and five-axis computer numerically controlled machine tools.[101] CASC's world-class simulation facility in western Beijing also aids cruise missile development by theoretically reducing testing requirements by 40-60 percent and shortening overall development time by 30-40 percent.[102] Acquisition of advanced Western systems could also reduce production time—Chinese engineers have reportedly had access to an intact Tomahawk that fell into Afghanistan territory in August 1998.[103]

Propulsion Systems. Engineers are also working on better propulsion systems that can increase the lethal range and/or speed of the cruise missile. Faster cruise missiles reduce an adversary's reaction time. In one of China's most significant aerospace programs, the PLA General Armament Department (GAD) and the Third Academy are designing a supersonic combustion ramjet engine (scramjet, or *chaoran chongya fadongji*) which can propel a missile at hypersonic speeds of between Mach 4-10.[104] Engineers are also working toward more efficient turbojet and turbofan engines and motors to significantly extend the range of cruise missiles. The anticipated range of China's first generation of land attack cruise missiles would be limited to around 300-400 kilometers. However, to be able to hit targets in Japan using a ground launched system, the Third Academy would have to produce a missile with a 1250-1500 kilometer range (750 kilometers for Okinawa).[105]

Radar Signature Reduction. With foreign assistance, China's defense industrial complex is also striving to reduce the radar cross-section of their cruise missiles.[106] The aerospace industry has produced radar absorbing material that targets the frequency range in which most acquisition radar operates (2-18 GHZ). While this material would not provide complete protection from radar detection, it could reduce the detection range of defensive radar. Engineers assert that radar-absorbing material, used in combination with contour shaping, can reduce the effective range of radar by 75 percent.[107] Radar absorbent materials and relatively high speed reduce warning time available to defenders and compress their timeline for detecting, tracking, identifying, and engaging the inbound missile.

Infrared Signature Reduction. Third Academy officials are also working to increase the survivability of their land attack cruise missiles by reducing their infrared (IR) signature. This could be achieved by the addition of an IR reduction tail cone which is designed to mix cool air that has traveled down the length of the airframe with hot air emitted from the jet exhaust. This would improve the survivability against IR sensors and IR homing missiles. Engineers are attempting to reduce the signature on cruise missile propellants, and conceal the location of mobile land attack cruise missiles.[108]

Other Terminal Guidance Systems. Other terminal sensor technologies under development include passive imaging infrared, CO2 laser radar, millimeter wave, and synthetic aperture radar terminal sensors, as well as various composite systems. R&D into passive imaging infrared sensors is focused on matching a stored computerized image with a real infrared image detected by the missile.[109] Third Academy engineers have already laid the technical foundation for a CO2 laser guidance system, to include the target recognition components.[110] Chinese aerospace engineers believe synthetic aperture radar, millimeter wave radar, infrared imaging, and laser radar guidance could result in an accuracy of one to three meters.[111]

THE CONVENTIONAL SECOND ARTILLERY

The PLA entity most responsible for deep strike missions against vital strategic and operational targets is the Second Artillery (*dierpaobing*). Since its establishment in the 1960s, the Second Artillery's mission has been limited to nuclear counterstrikes. Since the conclusion of the Gulf War, however, Chinese planners have diversified the Second Artillery's mission to include conventional strikes against high value strategic targets. The Second Artillery's adoption of a conventional strike role marks one of the most significant developments in PLA modernization. This discussion of the conventional Second Artillery outlines: 1) its organizational structure, 2) the vital role of information denial in Second Artillery operations, and 3) the conventional theater missile campaign doctrine and operations.

Organization.

The Second Artillery, with an estimated 90,000 personnel, consists of headquarters elements, six launch bases (*jidi*), one engineering design academy, four research institutes, two command academies, and possibly an early warning unit.[112] As key operational strike units, brigades are likely only assigned one type of missile to facilitate command and logistics. The Second Artillery headquarters and subordinate bases oversee warhead and missile storage facilities; maintenance units; and special warhead/missile transportation services.[113]

The 80302 Unit, headquartered in the mountain resort town of Huangshan, Jiangxi province, is the Second Artillery's most important base for conventional long-range precision strikes against Taiwan.[114] The Huangshan base includes both nuclear and conventionally armed theater missiles. During a wartime situation, multiple conventional brigades would be subsumed into a conventional theater missile corps (*juntuan*) consisting of a corps command post, a corps logistics command post, and a number of subordinate theater missile brigades each with different types of theater missiles. The corps command post would largely consist of command authorities from Beijing and Huangshan.[115]

The corps/base also oversees a set of "equipment assurance units" (*zhuangbei baozhang budui*) which includes a missile/warhead storage unit (*zhuangbei jishu qinwu budui*), a transfer station (*zhuanyunzhan*), and a repair depot (*tezhuang xiulicang*). Other corps/base support elements include a reconnaissance unit (*jizhen dadui*); a surveying/mapping unit (*cehui dadui*); a computer center (*jisuan zhongxin*); a weather center (*qixiang zhongxin*); a communications regiment (*tongxintuan*); an ECM regiment (*dianzi duikangtuan*); and an engineering regiment (*gongchengtuan*). Additional engineering, air defense, and anti-chemical units can be assigned as needed.[116]

A typical conventional theater missile brigade has a staff consisting of a headquarters, political, logistics, and equipment technology (*jizhuangbu*) departments. Brigade elements include a mobile brigade command post, a central depot (known as a "technical position" or *jishu zhendi*), a transfer point (*zhuanzai changping*), and an assigned set of pre-surveyed launch sites (*fashe zhendi*), as well as a set of reserve (*daiji*) launch sites. A conventional missile brigade also has a set of "equipment assurance sub-units" (*zhuangbei baozhang fendui*).[117] Brigades have at least four firing battalions (*fasheying*), with each battalion assigned at least three-four companies.[118] Companies subordinate to the launch battalion likely would be assigned at least one launcher, an electric power generation vehicle (*fadianche*), a surveying vehicle (*cekongche*), a communications command vehicle (*tongxun zhihuiche*), and a missile transport vehicle (*daodan yunshuche*). Battalions and companies would be assigned a zone within which to operate.[119]

Information Denial and the Theater Missile Campaign.

Key to the success of a theater missile campaign is concealing the forward deployment of brigade elements. Surprise can only be achieved through denial of foreign human and technical intelligence assets. To ensure a high degree of concealment, the Second Artillery has approached information denial in three ways: 1) communications security; 2) passive and active counterspace measures, and 3) a supporting space tracking network.

Secure Communications. Denying a potential adversary the ability to monitor communications and other electromagnetic emissions is a top priority. Beijing is examining a wide range of technologies to reduce the vulnerabilities of its communications to interception or jamming. Beijing has issued directives to strictly implement communications security (COMSEC) measures.[120] Introduction of fiber optics communications significantly increases its communications security. Engineers are studying the application of spread spectrum and frequency hopping technology for Beijing's satellite tracking and control network, as well as more secure satellite communications methodologies.[121] China is also

137

investing in more complex encryption (*mimaxue*) algorithms.[122]

Passive Counterspace Measures. The doctrinal requirement for preemption and secrecy is also leading the PRC toward development of passive and active counterspace measures. The PLA is stressing passive counterspace operations in an attempt to deny foreign reconnaissance satellites with information on its disposition of forces and R&D programs. Writings from the Academy of Military Sciences (AMS) indicate the PLA has a concerted effort to defeat opto-electronic, infrared, and radar reconnaissance systems. Specific measures include the exploitation of natural camouflage, and deception (*qipian*), to include distribution of false indicators and intelligence.[123] Chinese R&D into camouflage, concealment, and deception is explicitly intended to counter air and space-based reconnaissance platforms.[124] In 1992, COSTIND and CASC established camouflage standards for missile development in order to counter foreign optical, infrared, and radar satellite systems.[125] PLA engineers have also published technical papers on methods to reduce infrared signature of underground facilities.[126]

Another approach to countering space systems is through electronic countermeasures. The GAD and China's electronics industry appear to be developing a jammer to counter radar satellites.[127] PLA affiliated publications assert that China is capable of damaging optical reconnaissance satellites through the use of high-powered lasers.[128] Such measures would deny an adversary use of a satellite, but not destroy the platform itself, perhaps avoiding escalation of the conflict. As a side note, GAD and CASC are also moving toward fielding jammers intended to complicate use of communication satellites and NAVSTAR GPS.[129]

Active Counterspace Measures. The PRC is also examining more lethal measures to negate foreign satellites if necessary. Open source literature strongly suggests that a

Chinese direct ascent anti-satellite (ASAT) program may be in the model development stage in which the space industry is identifying various design proposals for seekers and propulsion systems. Chinese writings indicate R&D of ASAT systems is intended to discourage attacks on their own space systems. Technical papers demonstrate some of the greatest obstacles in developing an active counterspace capability are with development of a homing kill vehicle and associated terminal guidance. Specific systems under evaluation and simulation include infrared, radar, and impulse radar terminal guidance.[130] Chinese engineers have conducted studies to counter satellite decoys as well.[131]

Space Tracking Network. The key to passive and active counterspace operations is a space-tracking network that can monitor satellites passing overhead. China currently can detect and track most satellites with sufficient accuracy for targeting purposes.[132] The PLA is modernizing and expanding its tracking network, which is operated by the PLA General Armament Department's China Launch and Tracking Control General (CLTC). CLTC is adding overseas links in Chile and the South Pacific island of Kiribati, and has contracted with France for access to data from its space-tracking network.[133] China Academy of Sciences' astronomical observatories in Nanjing and Kunming feed into the CLTC network, providing orbital prediction data. CAS and CLTC are upgrading their network of high-resolution telescopes, augmented by laser tracking devices. China's space community claims an ability to detect objects in space down to 10 inches.[134] The CLTC space tracking network likely supports the Second Artillery through alert messages indicating that foreign reconnaissance satellites are passing overhead.[135]

The Phased Campaign.

A PLA theater missile campaign could take a number of forms. An initial option would be to use theater missiles as a show of force, similar to the missile exercises of 1995-96.

U.S. reaction to the last limited show of force, however, may have demonstrated that this option is not viable. If the PLA indeed desires to deny U.S. intervention through a *fait accompli*, a slow, gradual limited use option would permit a buildup of U.S. forces in the region. PLA operational concepts call for large scale, preemptive operations. Preemptive theater missile strikes, carried out in conjunction with air strikes and special operations, are intended to create favorable conditions for dominance in all dimensions of theater warfare.

A theater missile campaign would support achievement of the "three superiorities" (*sanquan*)—information dominance (*zhixinxiquan*); air superiority (*zhikongquan*); and sea superiority (*zhihaiquan*).[136] Strikes supporting the quest for information dominance would target the civilian and military leadership, semi-hardened command and control centers, weak links in Taiwan's defense information infrastructure, key intelligence collection facilities, and electronic warfare facilities. PLA conventional ballistic and land attack cruise missiles would attempt to paralyze (*tanhuan*) Taiwan's command and control system by cutting off fielded military forces from the civilian and military leadership in Taipei. Anti-radiation missiles would be employed against key radar installations.[137]

To achieve air superiority, the PLA would target key air defense sites and airfields. The PLA would seek to damage Taiwan Air Force runways, taxiways, weapons storage facilities, airfield command posts, and fuel depots to complicate generation of sorties. Strikes against airbase runways and taxiways are referred to as an "airbase blockade" (*fengsuo jichang*). The objective would be to shock and paralyze air defense systems to allow a window of opportunity for follow-on PLAAF strikes and rapid achievement of air superiority. Air superiority is key to establishing a no fly zone; enabling freedom of action on the ocean for a blockade; or permitting greater freedom of action for physical occupation of the island if necessary.[138]

To achieve sea superiority, PLA writings indicate prioritization of strikes against naval ports. The key objective will be to strike naval facilities in the opening phases of conflict as a means to prevent projection of naval power and resupply of strategic resources by sea. "Strike opportunities" exist when ships are concentrated in port or when they are moving along known transit routes en route to the theater of operations.[139]

Unsubstantiated reports indicate that a phased campaign could require at least 400 theater missiles distributed in as many as seven conventional missile brigades.[140] To maximize firepower for the most likely scenario, they most probably would be based in the Nanjing Military Region or chopped to the Taiwan theater of operations joint command during a crisis. PLA writings indicate that approximately 50 percent of its total theater missile inventory would be used in the initial strike phase. Western sources believe the PLA may deploy as many as 650 SRBMs opposite Taiwan over the next several years.[141]

The theater command center (*zhanyi zuozhan zhongxin*) would direct the missile campaign as one component of a joint strike force that also would include air forces, ground force artillery and tactical missiles, electronic attack assets, and special operations.[142] Coordination will be carried out via a firepower coordination cell (*huoli xietiaozu*) within the theater command center.[143] PLA officers envision a four phase theater missile campaign: 1) operational preparations phase (*zuozhan zhunbei jieduan*); 2) campaign mobility phase (*zhanyi jidong jieduan*); 3) missile strike phase (*daodan tuji jieduan*); and 4) enemy counterattack phase (*kangdi fanji jieduan*).[144]

Operational Preparations Phase.

After a CMC determination on the appropriate course of action (*juexin*), the operational preparation phase likely would include development or review of a mobility plan, increased security, and closer monitoring of foreign satellites and air/naval activity in the Western Pacific. Working in conjunction with the theater command, missile reconnaissance officers and planners probably would review or develop targeting folders. General Staff Department and theater intelligence staff would exploit existing intelligence and/or task space-based imaging assets for updates to support targeting. The firepower coordination cell within the theater command center would prioritize detected targets in keeping with the guidance of higher command for the conduct of the theater campaign and determine the most effective method of dealing with those targets. The theater command would de-conflict strikes so that firepower is not wasted, a complicated and time-consuming process. Theater commanders also would modify preplanned targeting of targets that have changed over time.[145]

Campaign Mobility Phase.

During the campaign mobility phase, brigade elements would deploy to the area of operations in a well-disguised fashion. Rail is the normal way of moving launchers and missiles from brigade garrison to a staging area or transfer assembly point (*zhuanzai changping*).[146] The individual launchers would then disperse to pre-surveyed launch sites (*zhendi*) within the battalion's assigned area of operations, not far from rail lines or highways.[147] A mobile command and control center would coordinate launches. Rapid reaction (*kuaisu fanying*) is essential, requiring a quick calculation of position, orienting the missile, inputting targeting data, and scattering in a very short period of time. Chinese writings indicate that units intend to launch within 40 minutes after arrival to the pre-surveyed launch sites.[148] To reduce reliance on pre-surveyed launch sites, however, the PLA appears to be integrating GPS onto their mobile launchers.[149]

Communication between firing units and upper echelons likely would be carried out through a mix of mobile

SATCOM, mobile digital microwave, and/or fiber optics. Due to its high level of security and reliability, the Second Artillery is trying to hardwire as much of its operational infrastructure as possible with fiber optics.[150] For security reasons, any wireless transmissions are to be limited to eight seconds or less. Operational orders would be transmitted through an automated command and control (C2) system due to the complexity and timeliness requirements of conventional theater missile operations. PLA officers note the requirement to integrate the Second Artillery's automated C2 system with that of the joint theater command's automated C2 system.[151]

Missile Strike Phase.

During the missile strike phase, Second Artillery units would support joint theater operations by striking strategic and operational centers of gravity. Missile firings would be coordinated with other strike assets and directed against critical nodes (*yaohai*) within an enemy's infrastructure. Chinese writings indicate that after an initial salvo, launchers could move to new pre-surveyed launch sites within that brigade's assigned area of operations.[152] At least three raids are feasible if one assumes availability of 400 theater missiles for the phased campaign.[153] The PLA intends to carry out synchronized launches from a wide range of azimuths in order to stress active missile defenses and associated battle management systems.[154] A range of space-based, airborne, and battlefield intelligence systems are needed to adjust firepower.[155]

The PLA has indicated prioritization of three target sets: 1) air and missile defense sites; 2) airfields and surface-to-surface missile sites; and 3) command, control, communications, computer, and intelligence (C4I) facilities. Neutralizing ground-based air defenses, airfields, and C4I facilities through multiple theater missile raids would present a window of opportunity for follow-on air strikes to consolidate air superiority over the island. PLA missile strikes against airfields could deny outside powers the ability to rush additional military equipment or military supplies to the island.[156] Some PLA-affiliated analysts speculate that parallel strikes against airfields, air defense sites, and other critical targets could permit PLA air superiority over the skies of Taiwan in as little as 45 minutes.[157]

Ground Based Air and Missile Defense. PLA writings identify ground-based air and missile defense units as primary targets. The critical node within an air or missile defense fire unit most likely would be its radar and command van. If no missile defenses existed, and CASC is able to meet the PLA's accuracy requirement of 20-45 meters, then only three to five missiles would be necessary to cause significant damage to key nodes within a fire unit with a high degree of confidence. To neutralize active theater missile defense units, PLA writings indicate use of coordinated strikes from multiple directions, using a combination of ballistic missiles, decoy drones, land attack cruise missiles, and/or anti-radiation missiles.[158] Radar and command vans could be subject to special operations attacks and electronic countermeasures. Because their re-entry speed precludes engagement by endo-atmospheric interceptors, conventional DF-21 MRBMs would be especially effective in neutralizing lower tier missile defense fire units. Guided submunition or an FAE payload likely would be the warheads of choice.[159]

Airfields and Surface-to-Surface Missile (SSM) Sites. Another critical target for PLA ballistic and land attack cruise missile strikes in a Taiwan scenario would be airfields and SSM sites. Senior Second Artillery officers write in internal journals that "attack opportunities" (*tuji de shiji*) will also exist against "intervening superpower" forces as they build up airpower in the region.[160] Airfields that could support offensive strike operations against the mainland would be the first priority. An "airfield blockade" would seek to damage runways, taxiway surfaces, and other critical nodes within an airbase. The PLA would need large numbers of theater missiles for a complete "airfield blockade." However, the PLA would need only a handful to impede Taiwan's ability to generate sorties. Strikes against runways likely would be particularly effective in temporarily pinning down much of the Taiwan Air Force.[161] Any runway damage would slow aircraft operations, simply because it takes time to determine the location and extent of the damage. Destruction of key facilities, such as airbase command centers, control towers, fuel depots, power generation facilities, and maintenance hangars would have a serious effect on air operations. Casualties to pilots and maintenance crews could be especially traumatic. Use of runway mines and targeting of unprotected rapid runway repair equipment would complicate recovery operations.[162] Warheads of choice for runway damage would include penetrator submunitions.[163]

To aid in its training, the PLA has constructed a mock-up of one of Taiwan's key airfields. The mock-up of Chingchuankang (CCK) airfield near Taichung includes an exact replica of the runway layout, taxiways, fuel storage, aircraft shelters, and revetments. The replica, located in a key training area in Gansu, 120 kilometers north of Jiayuguan, is intended for both theater missile exercises and air strikes.[164]

Leadership Facilities and C4I Centers. The PLA could strike at the heart of Taiwan's political and military leadership to impede the command and control of its forces. Early warning and technical intelligence collection sites could be subject to ballistic and anti-radiation missile strikes and electronic countermeasures. Political and military leadership facilities, such as the Presidential Palace and MND headquarters, are soft targets that would require fewer than five missiles to destroy each with a high degree of confidence. Fuel air explosive warheads are considered the optimal choice for strikes against softer political and military targets. Semi-hardened command and control and intelligence centers would require penetration warheads.[165]

Foreign Intervention. The PLA has indicated a willingness to use highly accurate SRBMs, MRBMs, and LACMs against U.S. assets, to include key bases in Japan and aircraft carriers operating in the Western Pacific. Chinese researchers have conducted extensive feasibility studies of the use of theater ballistic missiles against aircraft carriers. Analysts have noted how such a capability would require four components: ocean surveillance (*haiyang jianshi*); mid-course guidance (*zhongduan zhidao*); terminal guidance (*moduan zhidao*); and applicable control systems to maneuver the reentry vehicle to the target. PLA proponents have proposed the use of GPS for mid-course inertial corrections and the use of a millimeter wave seeker for terminal guidance.[166] Aware of the vulnerability of millimeter wave seekers to jamming, PLA engineers are surveying ECCM techniques to ensure effectiveness of terminally guided ballistic missiles.[167] In addition to aircraft carriers, Chinese writings indicate other targets would include regional airbases, naval facilities, and key C4I and logistical nodes.[168]

Counterstrike Phase.

For the counterstrike phase, PLA planners rely on survivability as a critical aspect of their theater missile force. In ensuring their survivability, designers believe three systems in particular pose the greatest challenges to the survivability of China's theater missile force: the F-117A, J-STARS, and AWACS. The most important step to ensure survivability is counter-reconnaissance (*fanzhencha*), that is, denying foreign air and space assets the ability to detect missile garrisons, storage facilities, and units in the field. Counter-reconnaissance measures include decoy launchers and missiles that must match the optical, infrared, and radar characteristics of real systems. The Second Artillery also intends to use natural masking, radiation reflectors, deception, and communications security. Chinese camouflage is explicitly intended to counter U.S. air and space-based reconnaissance platforms.[169]

There are indications that each theater missile brigade will have an organic electronic countermeasures regiment equipped with specially designed equipment which automatically activates an integrated system of radar jammers, lasers, chaff, flash bombs, and smoke. According to one report, the system was developed in large part to counter air-to-ground munitions delivered by aircraft such as the F-117A.[170]

CONCLUSION

A space-based surveillance architecture, the transition to a force structure dominated by theater missiles, and adoption of operational principles that stress preemption and surprise have serious implications for regional stability. An alleged arsenal of over 650 SRBMs—augmented by additional conventional MRBMs and LACMs—could provide Beijing with a conclusive edge in a future Taiwan Strait conflict. Such a force could also hold U.S. forces in the western Pacific at risk should a decision be made to intervene.

China's growing presence in space is intimately related to the PLA's emerging capacity for theater strike operations. Reconnaissance satellites are important for strategic and operational intelligence, indications and warning, and targeting. Space imagery is also needed to support battle damage assessments. Digitized satellite imagery is crucial for land attack cruise missile mission planning. In addition, space systems could enable the detection, tracking, and targeting of U.S. forward-deployed assets operating in the western Pacific Ocean. The same space-tracking network that manages China's space assets is crucial for operational security during a theater missile campaign.

Operational Implications.

China's emerging capacity for deep strike missions has a number of operational implications. First, theater missiles serve as critical enablers for dominance in other spheres of warfare. Of most significance is the relationship between theater missiles and the rapid achievement of air superiority. Consistent with emerging PLA doctrine of "rapid war, rapid resolution," a successful PLA theater missile campaign could strip Taiwan of its ability to effectively conduct air operations in a matter of hours (or minutes, according to PLA propaganda). Strikes against key air defense units and airfields would result in a temporary suspension of Taiwan air operations, creating a more permissive environment for PLA Air Force operations over the island. Air superiority, like the missile strikes, is not an end in itself. However, lessons absorbed from the Gulf War and the air campaign in Yugoslavia have demonstrated that air superiority enables other missions to take place with reduced costs and greater efficiency.[171]

Furthermore, theater missile operations could also quickly degrade Taiwan's capacity for naval warfare and ground operations. Fifty percent of the PLA's theater missile arsenal is to be dedicated toward the opening phase of conflict. Remaining missiles likely would be held in reserve to support naval and ground operations. Theater missile strikes against harbors and piers would complicate naval operations. Strikes against key bridges and staging areas would impede Taiwan Army counter-landing operations.

Also, China's expanding network of space sensors and long-range strike assets could pose a fundamental challenge to the U.S. ability to project power into the western Pacific Ocean. Increasingly accurate and lethal theater missiles could raise the costs of U.S. intervention in conflicts around the periphery of China. Space-based reconnaissance assets could facilitate detection of U.S. air and naval deployments into the area of operations. The PLA clearly understands U.S. vulnerabilities that arise from dependence on in-theater ports, airfields, logistics facilities, and C2 nodes. Successful fielding of terminally guided theater ballistic missiles could pose challenges to aircraft carrier battle groups, especially if operating within range of China's large inventory of 600-kilometer range SRBMs.[172]

Political Implications.

Developing a capacity for theater missile operations has political implications as well. Taiwan has enjoyed a defensive advantage over the mainland for many years. Adequate warning time and a robust defense has enabled Taiwan to blunt PLA air, naval, and ground assaults long enough to allow the international community to adjust to the situation, decide on a course of action, take diplomatic action, and/or flow forces to the region if necessary. However, a successful theater missile campaign—combined with information operations and air strikes—could enable Beijing to quickly strip Taiwan of its warfighting capacity.

To maintain the political and military viability of its new "trump card," Beijing has launched a coordinated foreign policy and propaganda campaign to shape the existing debate within the United States on defensive measures intended to counter theater missiles. Beijing generally poses six arguments against missile defenses, including an assertion that defenses will cause an arms race.[173] Beijing's campaign against missile defenses exploits biases by some within the United States against missile defenses. A mutually supporting dynamic exists between PRC officials and U.S. critics whose views on missile defenses are founded on traditional nuclear stability paradigms.

However, the Taiwan Strait case may be unique in that it is the first theater in which highly accurate *conventional* ballistic missiles dominate the strategic landscape. PRC officials, echoed by many within U.S. governmental and academic circles, argue that defenses against the growing PLA conventional theater missile threat would be destabilizing since they would spark an arms race.[174] However, a number of studies have demonstrated that in the conventional context, defenses generally have not been the causes of arms races. Conventional arms races are sparked or intensified by a rapid buildup of offensive capabilities.[175]

The misplaced focus on missile defenses within academic and policy communities in the United States has resulted in neglect of at least three dangers presented by Beijing's growing arsenal of increasingly accurate and lethal theater missiles. *First,* the conventional wisdom is that force would only be used against Taiwan in the event the government legally declares the island as an independent political entity. However, overwhelming offensive capabilities increase the chances that force could be used short of a *de jure* declaration of independence. Confidence in a quick military victory could lower the perceived cost of conflict and thus increase Beijing's incentives to use force. At a minimum, a decisive PLA advantage in offensive capabilities would increase risks of greater PRC bellicose behavior in the cross-Strait relationship. In addition, the ability to strip Taiwan of its capacity for military operations—in effect a first strike capability—raises dangers of preemptive war.[176] The PLA preemptive strike doctrine is also destabilizing since it decreases warning time that could allow for diplomatic intervention. An overwhelming offensive advantage may also reduce Beijing's incentives for arms control and confidence building measures, and reduce their willingness to compromise in future cross-Strait dialogue.

Second, reduced costs for military action could lead to another unintended consequence of the theater missile buildup—a Taiwan punitive deterrent to raise the costs of PLA military action. At least one punitive deterrent is Taiwan's own theater missile capability. A Taiwan ballistic or land attack cruise missile would serve as a political tool to raise the costs of PRC military action. Even more ominous is that a severe collapse in their sense of security could prompt Taiwan to renew efforts to develop a nuclear device. Taiwan is considered by some to have the capacity to develop nuclear weapons quickly if the need should arise. Within the last 2 years, there has been an open debate in Taiwan regarding the utility of developing weapons of mass destruction.[177]

Third, as Taiwan's national security community debates the need for a deterrent, the magnitude of the theater missile challenge may increase domestic pressure for tactically offensive counterforce operations, to include preemptive strikes. Theoretical studies have demonstrated that maintenance of an exclusively defensive force posture against an overwhelming offensive force is prohibitively expensive.[178] Tactical offenses in support of a strategically defensive doctrine are more cost effective. As the PLA theater missile threat evolves, Taiwan strategists may adopt operational concepts outlined in U.S. Department of Defense *Joint Pub 3-01.5* that states "the preferred method of countering enemy theater missile operations is to attack and destroy or disrupt theater missiles prior to their launch." This notion comes as no surprise to the PLA Second Artillery, an organization whose doctrine rests on the assumption that their phased campaign would be answered with Taiwan or U.S. counterattacks.

Theater Missile Countermeasures.

A preemptive strategy that relies on an overwhelming offensive force is not only destabilizing, but may be risky from a warfighting perspective. The outlook described above is admittedly pessimistic and worst-case. The posited aim of a PLA air and missile campaign is strategic paralysis, with the expectation being that "paralysis" must somehow equate to "surrender." It may not work that way. With proper preparations, Taiwan, or any other adversary, could recover from initial attacks. Observers have asserted that Taipei would fold after the impact of a single missile on Taiwan. However, lessons from World War II, the Vietnam

War, and elsewhere have shown that strategic attacks could harden rather than diminish resolve.[179]

A number of steps could be taken to reduce the operational effectiveness of the PLA theater missiles and supporting surveillance assets. The theater missile problem is already forcing the Taiwan military to modernize in a way that they would not have otherwise. The only way to effectively counter a large-scale theater missile threat is through jointness and innovative warfighting concepts commonly associated with the RMA. Assuming requisite changes and investments are made, the PLA's ability to achieve a decisive victory over Taiwan is not assured.

Perhaps the most important countermeasure is a survivable C4I architecture and robust passive defenses. Passive defense includes: 1) tactical warning; 2) reducing the effectiveness of PLA targeting through operational security, deception, and mobility; 3) reducing vulnerability through hardening, redundancy and robustness, dispersal, and effective civil defense; and 4) recovery and reconstitution. In addition, the PLA's successful fielding of sophisticated terminal guidance systems would be accompanied by a new set of vulnerabilities. GPS, and optical, radar, and millimeter wave seekers can be jammed, as could the PLA's future space reconnaissance assets.[180]

Furthermore, the complexity of a theater missile campaign presents opportunities for "induced friction." The challenges presented by an overwhelming capacity for offensive operations would naturally prompt defenders to prevent the launch of theater missiles. This would be carried out by attacking elements of the overall system, including such actions as destroying launch platforms, reconnaissance, surveillance, and targeting platforms, command and control nodes, missile stocks, and transport infrastructure. Strikes against selected nodes in a theater missile brigade could have significant systemic effects that could reduce the frequency or intensity of theater missile strikes.

The effectiveness of theater ballistic and land attack cruise missiles strikes could also be reduced through active missile defenses. Exclusive reliance on active defenses, however, would be cost prohibitive and only partially effective against the type of theater missile threat that Taiwan is expected to face. The most serious challenge to active defenses may be the tyranny of geography—Taiwan is situated close enough to the mainland to allow the PLA to launch from a wide range of azimuths. Multi-axis theater ballistic missile attacks could stress even the best battle management and command, control, and communication systems, especially if combined with air and LACM strikes, electronic attack, and special operations.[181]

In the end, however, the optimal solution lies in creating incentives for Beijing to moderate its theater missile deployments. The first step is recognizing the destabilizing nature of the PLA theater missile buildup. While urging PLA restraint in deploying theater missiles opposite Taiwan is a worthwhile endeavor, one should not be overly sanguine about the chances for success. Theater missiles are an integral part of the PLA's overall modernization objectives. As long as the PLA seeks to develop the kind of force that could give the PRC a decisive military advantage over Taiwan, then the ability to freeze or roll back theater missile deployments will be limited. Nevertheless, greater effort must be made to convince the civilian leadership in Beijing that the large-scale deployment of offensive weapons would adversely affect regional stability and that resolution of sovereignty disputes through other than peaceful means is not a viable option.

CHAPTER 5 - ENDNOTES

1. According to Department of Defense *Joint Publication 3-01.5*, "*theater missile* applies to ballistic missiles, cruise missiles, and air-to-surface missiles whose targets are within a given theater of operation." This term generally does not apply to shorter-range systems such as Maverick and Harpoon.

2. Wang Xuejin and Zhang Huaibi, "Didi changgui daodan budui zuozhan zhidao sixiang fenxi," (Analysis of Conventional Surface-to-Surface Missile Operations Guiding Thought) in *Lianhe zhanyi yu junbingzhong zuozhan* (Joint Theater and Service Operations), Beijing: National Defense University Press, 1998, pp. 223-227.

3. Nan Shih-yin, "Inside Story of Enlarged Central Committee Meeting," Hong Kong *Kuang chiao ching*, January 16, 1996, in *Foreign Broadcast Information Service-China (FBIS-CHI)*-96-027; also see Jen Hui-wen, "Latest Trends in China's Military Revolution," in Hong Kong *Hsin Pao* (Hong Kong *Economic Journal*), February 9, 1996, in *FBIS-CHI*-96-047; for other comments on lessons from the Gulf War, see Ho Po-shih, "The Chinese Military Is Worried About Lagging Behind in Armament," *Tangdai*, March 9, 1991, pp. 17-18.

4. Liang Zhenxing, "New Military Revolution and Information Warfare," *Zhongguo dianzi bao (China Electronic News)*, October 24, 1997, p. 8, in *FBIS-CHI*-98-012.

5. According to *Joint Pub 1-02*, air superiority is that degree of dominance in the air battle of one force over another which permits the conduct of operations by the former and its related land, sea, and air forces at a given time and place without prohibitive interference by the opposing force.

6. For one of the best overviews of these doctrinal shifts, see Nan Li, "The PLA's Evolving Warfighting Doctrine, Strategy and Tactics, 1985-95: A Chinese Perspective," in *China Quarterly*, July 1995, pp. 443-463; and Nan Li, "The PLA's Evolving Campaign Doctrine and Strategies," in James C. Mulvenon and Richard H. Yang, *The People's Liberation Army in the Information Age*, RAND: Washington, DC, 1999, pp. 146-172. For more detail, see Liu Mingtao and Yang Chengjun, *Gaojishu zhanzhengzhong de daodanzhan*, (*Missile War Under High-Tech Conditions*), Beijing: NDU Press, 1993, pp. 5-26; also see Li Qingshan, *Xin junshi geming yu gaojishu zhanzheng (New Military Revolution and High Tech Warfare)*, Beijing: AMS Press, 1995; Liu Senshan and Jiang Fangran, *Gaojishu jubu zhanzheng tiaojianxia de zuozhan (Operations Under High Tech Local War Conditions)*, Beijing: AMS Press, 1994, pp. 13-33; and Senior Colonels Huang Xing and Zuo Quandian, "Operational Doctrine for High Tech Conditions," *Zhongguo junshi kexue (China Military Science)*, November 20, 1996, pp. 49-56, in *FBIS-CHI*-97-114.

7. Senior Colonel Jiang Lei, *Xiandai yilie shengyou zhanlue (Modern Strategy of Pitting the Inferior Against the Superiority)*, Beijing: NDU

Press, pp. 6-49. Colonel Jiang is one of the few PLA officers awarded a Ph.D. in Operations Research from AMS. He is currently assigned to AMS Strategic Studies Department; on the pre-emptive strike concept, see Lu Linzhi, "Pre-emptive Strikes Endorsed for Limited High Tech War," *Jiefangjun Bao*, February 14, 1996, in *FBIS-CHI*-96-025. Among numerous references, see, for example, Wang Pufeng, *Xinxi zhanzheng yu junshi geming* (*Information Warfare and the Military Revolution*), Beijing: AMS Press, 1995.

8. With air and sea superiority, some of more seemingly outlandish scenarios for an amphibious invasion, such as the large-scale use of PLA commercial fishing vessels, become more feasible.

9. For a summary of Chinese writings on perceived U.S. weaknesses, see Dr. Michael Pillsbury, *Dangerous Chinese Misperceptions: The Implications for DoD*, Prepared for Office of Net Assessment, 1998.

10. See Desmond Ball, "Signals Intelligence in China," *Jane's Intelligence Review*, August 1, 1995, pp. 365-375; and Robert Karniol, "China Sets Up Border SIGINT Bases In Laos," *Janes Defense Weekly*, November 19, 1994, p. 5.

11. China Aerospace S&T Corporation (CASC), directed by Wang Liheng, is an offshoot of the former China Aerospace Corporation. The new organization adopted the former's CALT (1st Academy), the 4th Academy, CAST (5th Academy), SAST (8th Academy), 062 Base, and the 067 Base. The China Aerospace Electromechanical Group (*Hangtian jidian jituan*) includes the Second and Third Academies and the 61 and 66 Bases.

12. There seems to be a debate within the PLA on control of China's future space architecture, pitting GAD against the Second Artillery. See Senior Colonel Ping Fan and Captain Li Qi, "A Theoretical Discussion of Several Matters Involved in the Development of Military Space Forces," *Zhongguo junshi kexue*, May 20, 1997, pp. 127-131, in *FBIS-CHI*-97-0302.

13. This conclusion is drawn from ONI's 1994 study, "Chinese Space-Based Remote Sensing Programs and Ground Based Processing Capabilities," discussed in Jefferey Richelson, ""Navy Says China Poised To Close Space-Intel Gap," *Defense Week*, February 24, 1997, p. 9.

14. *Ibid*. Weather satellites are also an important sensor but are beyond the scope of this study.

15. Data collected by China's airborne SAR remote sensing platform can be transmitted real time to a ground station that is within 120 kilometers of the aircraft. A tactical ground processing system equipped with a VSAT terminal can then transmit the data to a command center. See "Remote Sensing Technical Systems for Reducing Flood Disasters," *Yaogan kexue xin jinzhan*, April 1995, in *FBIS-CST*-96-002; and Xu Guanhua and Guo Huadong, "Progress, Mission of Remote Sensing Research," *Yaogan kexue xin jinzhan* (*New Progress of Remote Sensing Science*), April 1995, in *FBIS-CST*-96-002.

16. The HY-1 is a small satellite which will based upon the CAST986 bus. The CAST986, an inexpensive common-use bus, will be used for a range of other satellites to include the SJ-5 experimental satellite, a constellation of electro-optical satellites, a generation-after-next navigation satellite constellation, and a tactical satellite communications (SATCOM) system. The 863 Program includes a special budgeting mechanism for R&D in seven key technology areas. The PLA is responsible for oversight of space and laser components of the 863 Program. The specific designation of the national SAR effort is the 863-308 program that also includes a near real-time electro-optical satellite system. The airborne system was tested on board a U.S. space shuttle mission (SIR-C). See Wang Wei, "State S&T Organs Approve Design of Spaceborne Synthetic Aperture Radar," *Zhongguo kexue bao* (*China Science News*), May 3, 1995, in *FBIS-CST*-95-010; "Woguo xingzai hecheng kongjing leida yingyong yanjiu qude zhongda jinzhan (*China's Satellite SAR Applied Research Achieves Tremendous Advances*), *Zhongguo hangtian*, February 1996, p. 16; "Spaceborne-SAR Modern Information Technology Highlighted," *Zhongguo kexue bao*, September 20, 1996, p. 4, in *FBIS-CST*-96-020; Yuan Xiaokang, "High Speed Data Transmission of Satellite-borne SAR," *Zhidao yu yinxin* (*Guidance and Detonators*), 1995 (4), pp. 8-14. 509th RI is summarized in *China Astronautics and Missilery Abstracts* (*CAMA*), Vol. 4, No. 4. Also see Yuan Xiaokang, "Performance Parameters and Design Requirements of Satellite SAR," *in Shanghai hangtian*, 1996 (3), pp. 12-18; and Long Zhihao. "Application of Radar Satellites," *Aerospace China*, November 1991, p. 29. Li Yudong, "Satellite-borne Radar Reconnaissance," *New Electronic Warfare Technology and Intelligence Reform Studies Abstracts,* 1995, 10, pp. 126-133. Li is from the Southwest Institute of Electronic Equipment (SWIEE). For comments on preliminary research on the second generation SAR satellite, see "China's Microwave-Imaging Radar Systems Engineering Highlighted," *Zhongguo kexue bao*, September 20, 1996, p. 4, in *FBIS-CST*-96-020.

17. Long Zhihao, "Leida weixing de yingyong" (Applications of Radar Satellites), *Zhongguo hangtian*, November 1991, pp. 29-31;

Zhang Wanzeng, "Weixing hecheng kongjing chengxiang leida de tedian jiqi zai junshi zhenchazhong de yingyong" (Applications and Characteristics of Satellite SAR for Military Reconnaissance), *Zhongguo hangtian*, November 1993, pp. 20-22. Zhang Wanzeng is assigned to the PLA GSD Second Department's Techology Bureau. Huang Weigen, Zhou Changbao, and Wan Zhongling, "Woguo xingzai SAR haiyang yingyong de xianzhuang yu xuqiu," (Current State and Requirements of China's Satellite-borne SAR for Marchitime Applications), in *Zhongguo hangtian*, December 1997, pp. 5-9. China began exploration of space-based SAR systems for antisubmarine warfare purposes in the 8th Five-Year Program (1991-1996). As a side note, a U.S. Los Alamos employee under contract for TRW was arrested in May 1999 for providing the Chinese information on a classified project he was working on with regards to SAR satellite imaging of submarines.

18. See Stokes, *China Strategic Modernization,* p. 38-39.

19. *Ibid*, p 22.

20. The United States and the Soviet Union attained a near-real-time capability in 1976 and 1982, respectively.

21. Jean Etienne, "Les Nouveaux Projets de L'Asie Spatiale," in *Space News*, No. 110, November 4, 1996, at *http://www.sat-net.com/space-news*. Also see Chou Kuan-wu, "China's Reconnaissance Satellites," *Kuang chiao ching*, March 16, 1998, pp. 36-40, in *FBIS-CHI*-98-098.

22. The 863 Program, specifically the 863-308 project, funds China's remote sensing program at least in part. Hong Mei, "Tactical Application Satellite Imagery System," *Hangtian fanhui yu yaogan (Spacecraft Recovery and Remote Sensing)*, Vol. 16, No. 1, 1995, pp. 30-33, in *CAMA*, Vol. 2, No. 5. A 700-kilometer orbit optimizes coverage at the expense of resolution—a lower orbit naturally will increase the resolution. See "China To Launch Ten More Satellites by 2000," *Xinhua*, February 22, 1998, in *FBIS-CHI*-98-053.

23. These concepts have been closely examined and strongly advocated by the space and missile industry. See Zhang Dexiong, "Guowai xiaoxing weixing de guti huojian tuijin xitong" (Solid Rocket Propulsion Systems for Foreign Small Satellites), in *Hangtian qingbao yanjiu*, HQ-93011, pp. 139-155; Wang Zheng, "Screening Studies and Technology for All-Solid Space Launch Vehicles," *Guti huojian fadongji sheji yu yanjiu (*Solid Rocket Engine Design and Research), April 1996, pp. 63-73, in *CAMA*, 1996, Vol. 3, No. 6; and Zhang Song, "Design and

Optimization of Solid Launch Vehicle Trajectory," *Guti huojian jishu*, Vol. 20, No. 1, 1997, pp. 1-5; and Zhang Dexiong, "China's Development Concept for Small Solid Launch Vehicles," CASC Fourth Academy Information Research Reports, the Fourth Edition, October 1995, pp. 1-11, in *CAMA*, Vol. 5, No. 2.

24. Mao Genwang and Wang Liang, "Weixing de junshi yingyong tedian, fazhan xianshi yu yingyong qianjing" (Military Satellites and their Prospects for Development), *Zhongguo hangtian*, May 1992, pp. 33-53.

25. Jefferey Richelson, "Navy Says China Poised To Close Space-Intel Gap," *Defense Week*, February 24, 1997, p. 9.

26. Mei Lin, "PLA Methods of Operations Assessed," *Chung kung yen chiu*, November 15, 1997, No. 371, pp. 50-60, in *FBIS-CHI*, March 10, 1998.

27. Chang Jijun, "Remote Sensing Image Data Compression and Its Performance Evaluation," *Kongjian jishu qingbao yanjiu*, July 1994, pp. 37-54.

28. "Woguo jiang zengshou wuke weixing de guance shouju" (China Will Receive Five Sources of Surveying Data), *China Space News*, January 5, 1997, p. 1; Fan Xizhe, "Future Importance of Information," *Zhongguo kexue bao*, December 16, 1996, in *FBIS-CST*-97-003. Also see announcement by SSTC Vice Minister Xu Guanhua, *Xinhua News Release*, in *FBIS-CHI*-96-247. Twelve subscribers will be able to link into the stored images. Also see Chiu Fangying, "China's Remote Sensing Image Digitization Equipment Meets Advanced International Standards," *Keji Ribao*, October 26, 1995, p. 1, in *FBIS-CST*-96-002.

29."Wo weixing yaogan tuxiang shuzihua shebei shijie lingxian" (Our Satellite Remote Sensing Digitized Imagery Equipment Leads the World), *Zhongguo hangtian*, January 1996, p. 39; also see "China's Satellite Remote Sensing Image Digitization Equipment Meets Advanced International Standards," *Keji ribao*, October 26, 1995, in *FBIS-CST*-96-002.

30. "China To Use Italian Software to Interpret Imagery," *Space News*, March 2-8, 1992, p. 23. Two other Chinese organizations involved in the project include China's Research Institute for Surveying and Mapping and the National Laboratory of Resources and Environmental Information Systems. Peng Yiqi, a senior engineer at the National Remote Sensing Center, led the Chinese negotiations.

31. China is working toward is the development of data relay satellites (*shuju zhongji weixing*). China signed agreements with France (1993) and Chile (1994) for joint use of their ground stations. Seeking to transmit imagery directly to theater and field commanders, China's remote sensing community has also begun to explore development of mobile remote sensing ground stations. On China's data relay satellite program, see Zhang Wanbin, "Spaceflight Development Strategy: Mid-Long Term Development Strategy," *Zhongguo keji luntan (Forum on Science and Technology)*, November 1992, pp. 9-12, in *JPRS-CST*-93-002; and Cheng Yuejin, "Information Transmission System of Data Relay Satellites," *Space Technology Information Research*, 1993 in *CAMA*, 1994, Vol. 1, No. 6. On China's contracting for access to French, Kiribat, and Chilean ground stations, see Wang Chunyuan, *China's Space Industry and Its Strategy of International Cooperation*, Stanford University, July 1996, p. 4. LTC Wang serves on the senior staff of COSTIND's foreign affairs bureau. On mobile ground station acquisition, see Wang Mingyuan, "Mobile Remote Sensing Ground Stations," *Kongjian dianzi jishu (Space Electronic Technology)*, 1997 (2), pp. 32-37 in *CAMA*, 1997, Vol. 4, No. 6. One of the first U.S. mobile imagery ground stations, EAGLE VISION, entered the inventory in 1995. China has expressed interest in acquiring a foreign EAGLE VISION-like system from either U.S. or French vendors.

32. *Dragon in Space*, July 24, 1999. Its data rate is about 150-300Mbps. SWIET is the major tracking and telemetry systems provider for the Chinese space program.

33. As of 1997, the State space budget was around $100 million, or 0.035 percent of the overall GNP.

34. "Luan Enjie fuzongjingli tan hangtian zhiliang" (Vice-General Manager Luan Enjie Discusses Space Quality), *Zhongguo hangtian bao (China Space News)*, March 21, 1994, p. 1.

35. China's space community is assessed to have grossed more than $500 million since the first commercial launch in 1990. China signed agreements with Russia for cooperative development in 10 areas, including surveillance systems, propulsion, joint design efforts, scientific personnel exchanges, space systems testing, and satellite navigation systems. See "Wang liheng fujuzhang lutuan fangwen e'wu liangguo" (CASC Deputy Director Wang Liheng Leads Delegation to Russian and Ukraine), *Zhongguo hangtian bao,* April 11, 1994, p. 1. Cooperation with France is focused on small satellite development, space tracking, and attitude control systems.

36. *Ibid.* Mei Lin, "New PLA Methods of Operations Assessed," *Chung Kung Yen Chiu*, November 15, 1997, No. 371, pp. 50-60, in *FBIS-CHI*, March 10, 1998.

37. PLAAF deficiencies are discussed at length in Kenneth W. Allen, Glenn Krumel, and Jonathon Pollack, *China's Air Force Enters the 21st Century*, RAND Project Air Force study, 1995, pp. 112-113.

38. Yuan Jun, "Zhanshu dandao daodan weixie yu fangyu de jiben wenti," (Fundamental Problems Associated With Tactical Ballistic Missile Threats and Defense), *Zhongguo hangtian*, November 1998, pp. 35-40.

39. Liu Mingtao and Yang Chengjun, *Gaojishu zhanzhengzhong de daodanzhan* (*Missile War In High Tech Warfare*), Beijing: National Defense University Press, 1993, pp. 4-26; and Wang Jixiang, "Inspiration for Chinese Ballistic Missile Development From the Gulf War," *Hangtian keji qingbao yanjiu baogao xilie wenzhai*, April 1994, pp. 49-56, in *CAMA*, Vol. 3, No. 6.

40. *Report to Congress Pursuant to Section 1305 of the FY97 National Defense Authorization Act*, p. 4. The report states that most of these missiles are likely to be short- or medium-range systems.

41. The $500,000 per missile figure is from Yuen Lin, "Probing the Capability of Taiwan's Antiballistic Missiles," Hong Kong *Kuang chiao ching*, August 16, 1998, pp. 54-61, in *FBIS-CHI*-98-252. In comparison, the cost of three MADS batteries with 180 missiles amounts to $850 million. An AEGIS destroyer runs around $850 million-$1 billion. Taiwan's FY98 defense budget totaled NT $275 billion (U.S. $916 million), amounting to 22.43 percent of the national budget. It is important to note, however, that the value of a missile defense system is judged on the basis of what is being defended rather than the costs of the offensive missiles.

42. It should be noted that recent Beijing-affiliated publications out of Hong Kong have resurrected the issue of the DF-25, a 2,000-kilometer range system that is armed with a 1,700-kilogram conventional warhead. The DF-25 allegedly utilizes the first two stages of the DF-31 ICBM. However, author's discussions in Beijing during the 1993-94 timeframe indicated that the DF-25 program had been cancelled in favor of the conventional DF-21. The primary difference between the two systems was the warhead size—the nuclear DF-21 only had a throw weight of 600 kilograms while the DF-25 was designed to have a 2,000-kilogram warhead. The DF-25 was first discussed in John Wilson Lewis and Hua Di, "China's Ballistic Missile Programs," *International*

Security, Fall 1992, pp. 5-40. If based at Tonghua (80301 Unit), the DF-25's 1700-kilometer range would have permitted strikes against the main islands of Japan, but not Okinawa. Assigning the 1700-kilometer system under the Luoyang base (80304 Unit) would have enabled strikes against all of Taiwan. Lewis and Hua asserted that the DF-25 was intended to defend claims in the South China Sea. However, to range the Spratly Islands, the system would have had to be based on Hainan Island. Past PLA deployment practices indicate that deployment of theater missiles on Hainan Island is not likely since: a) theater missile units are unlikely to stray too far from their established base headquarters; and b) basing on Hainan renders the units vulnerable to strikes.

43. Duncan Lennox, ed., *Jane's Strategic Weapon Systems,* Issue 24, May 1997, Surrey, England: Jane's Information Group.

44. See Edward R. Harshberger, *Long Range Conventional Missiles: Issues for Near-Term Development*, RAND: Santa Monica, 1991, p. 142. For an upper range estimate of 60 meters, see Lin Chien-hua, "What Equipment Should Taiwan Use to Defense Itself," *Taipei Tzu-li Wan-pao*, November 9, 1997, p. 2, in *FBIS-CHI*-97-364.

45. George Lindsey, *The Information Requirements for Aerospace Defense: Limits Imposed by Geometry and Technology*, Bailrigg Memorandum 27, CDISS, Lancaster University, p. 18. If moved closer to the target, the DF-15 likely would be launched on a lofted trajectory that would increase the flight time outside the atmosphere, thus increasing the missile's vulnerability to upper tier systems. On the other hand, a lofted trajectory could increase the missile's reentry speed, reducing the footprint, or defended area, of lower tier systems such as PATRIOT.

46. Zhao Yunshan, *Zhongguo daodan jiqi zhanlue, jiefangjun de hexin wuqi* (*China's Missiles and Strategy: The PLA's Central Weapon*), Hong Kong: Mirror Books, p. 232. Other sources credit the DF-15 with only as good as a 150-meter CEP. See "Missiles! China Has Them Too!" *Wen wei po*, June 1, 1999, p. A5, in *FBIS-CHI*-00169, June 22, 1999.

47. *Ibid.* Informed sources assert the Mirror (*Mingjing*) series of books have a mixed record of reliability. Zhao states that the expanded range DF-15 incorporates a more advanced propellant. There is often confusing reporting on an unidentified 1000 kilometer system—the M-18—that may in fact be the rumored extended range DF-15. While an extended range DF-15 can not be confirmed, there certainly could be a motive for developing a conventional theater ballistic missile with a 1,200-kilometer range. First, a 1,200-kilometer range system would

significantly reduce the defended area or "footprint" of land- and sea-based lower tier missile defense systems due to its reentry speed. Because of its existing infrastructure, one could speculate that an extended range DF-15 brigade could be established under the Second Artillery's Huaihua Base (80305 Unit). Huaihua, situated in western Hunan Province, is just over 1,200 kilometers from Taipei. Secondly, a 1,200-kilometer DF-15 fired from a notional site in the area of Nanping in Fujian province could easily range Kadena AB, Okinawa, and all of the Luzon Strait (Bashi Channel).

48. See the 1999 *DoD Report to Congress on the Security Situation in the Taiwan Strait*. It should be noted, however, that foreign sources familiar with the PLA believe that the 300-kilometer DF-11 has already been fielded by at least two PLA ground force group armies. In addition, a March 29, 1999, edition of *Jiefangjun bao* discusses the conversion of an unidentified Nanjing Military Region artillery unit to an SRBM brigade. The conversion began in early 1997. The author is indebted to Ken Allen for this information.

49. Zhao, p. 234.

50. *Report to Congress on Theater Missile Defense Options in the Asia-Pacific Region*, February 1999.

51. The 700-meter CEP is extracted from *Jane's Strategic Weapons Systems,* 1998. The conversion of the DF-21 from a strictly nuclear mission to a conventional role was reported as early as 1994 in the Chinese journal, *Guoji Hangkong (International Aviation)*. Initial indications of a terminally guided DF-21 are from discussions between Richard Fisher, who was a Senior Policy Analyst at the Heritage Foundation, and an engineer from CALT's Beijing Research Institute of Telemetry (704th Research Institute) at the 1996 Zhuhai Air Show. Extensive CASC technical writings on terminally guided theater ballistic missiles tend to substantiate the engineer's comments. Other sources indicate that the conventional DF-21 program entered the applied R&D phase in 1995 (*xinghao yanzhi*) and that the primary payload will be a penetrator warhead (*zuandi dantou*) for use against semi-hardened facilities such as command centers. A radar terminal guidance package will enable a 50-meter CEP. The conventional DF-21 range could be extended to 2200 kilometers. See Will Young, "Shenmi de zhongguo daodan budui" (The Development of the Chinese Second Artillery), *Shijie junshi luntan (World Military Forum)*, Internet edition (*http//:www.wforum.com*), January 2000, in Chinese.

52. Zhu Bao, "Di-di dandaoshi zhanshi daodan de fazhan qushi," pp. 9-19. The CEP is the radius of a circle within which 50 percent of missile fired will impact.

53. John Wilson Lewis and Hua Di, "China's Ballistic Missile Programs: Technologies, Strategies, and Goals," *International Security*, Vol. 17, No. 2, Fall 1992, p. 29.

54. Zhu Bao, pp. 9-19. Development of VLSIC and LSIC technology is one of Beijing's highest priorities. In one effort, China has invested RMB 1.39 in the Huajing Group's Project 908 program, which seeks to miniaturize and mass-produce VLSIC/LSICs. China hopes to develop sub-micron VLSICs in the next few years. See Zhang Longquan, "Huajing Group Builds 'Project 908' VSLIC Production Line," *Jisuanji shijie*, January 8, 1996, No. 2, p.1, in *FBIS-CHI*, January 8, 1996. With the help of Project 908, CASC's Ninth Academy would likely actually produce the application-specific integrated circuits. The SCUD-B payload known as the AEROFON uses an optical sensor during the latter stages of flight to detect and home in on a target.

55. Xie Lei, "Technical Research on Millimeter Wave Guidance," *Aerospace S&T Information Studies Series Abstracts* (6), pp. 235-250; and Xie Lei, "Application of Millimeter Wave and Infrared Technologies in Weapons Systems," *Aerospace S&T Information Studies Series Abstracts* (7), pp. 241-258.

56. Zhu Bao, pp. 9-19.

57. For a summary of test results, see Sun Mei, "GPS For Evaluating Inertial Measurement Unit Errors," *Hangtian kongzhi (Aerospace Control)*, 1995, Vol. 13, No. 1, pp. 69-75, in *CAMA*, 1995, Vol. 2, No. 4; also see Wang Shuren, "Principles of Onboard GPS Navigation Transponders," *Hangkong dianzi jishu*, undated, pp. 20-23. Wang is from the Second Artillery's Academy of Engineering.

58. Li Yonghong, "Ballistic Trajectory Determination Using the Differential Global Positioning System," *Binggong xuebao*, 1997, 18 (4), pp. 372-374. DGPS upgrades the civilian GPS signal though a precisely located GPS station that broadcasts a correction signal on a different frequency to other GPS receivers. In addition to military uses, a DGPS reference station is often used for surveying and maritime safety. Reference updates can be transmitted to the missile via a data link. As part of its Ninth 5-year Plan, China is constructing 20 DGPS stations along its eastern coast, each with a range of 300 kilometers. The positioning accuracy is five meters, a marked improvement from the original positioning system's minimum error of 100 meters.

59. See, for example, Bill Gertz, *Betrayal*, Boston: Regnery Press, p. 249.

60. Zhu Bao, "Di-di dandaoshi zhanshi daodan de fazhan qushi" (Developmental Prospects of Surface-to-Surface Tactical Ballistic Missiles), pp. 9-19.

61. *Lianhe zhanyi di erpaobing zuozhan (PLA Second Artillery Joint Campaign Operations)*, unpublished manuscript, 1996, p. 11. The document is believed to be an internal PLA academic paper, but its authenticity has not been established. However, a number of sources have corroborated much of the paper's content. For technical studies, see Yang Xiaolong, Wan Chunxi, and Li Xingcheng, "General Technical Research on Use of Strategic Missile Terminal Submunitions for Blocking Airbases," Space/Missile General Information Network Conference Paper (97021), October 1997; Yu Renshun, Qi Zhanyuan, Yang Xiaolong, "Guidance Law of Terminally Guided Submunitions for Attacking Runways," *Zhanshu daodan jishu*, February 1998, pp. 25-31. Authors are from Bejing Ligong Daxue. For a study addressing submunition dispersal problems, see Yan Dongsheng, "Technical Means for Reducing Dispersal of Mini-Warheads," paper presented at the October 1995 Annual Conference on Flight Mechanics. Xu is from CALT's 14th Research Institute, the entity responsible for warhead development. For other studies on use of missiles against airfields, see Yu Renshun, "Research on Terminally Guided Submunitions For Blocking Airfield Runways," paper presented at November 1997 conference of National Missile Designers Network, in *CAMA*, Vol. 5, No. 3; Yang Bingwei, "Structural Design Problems and Test Methods of Anti-Runway Penetrators," in *Aerospace S&T Intelligence Studies Abstracts (5)*, 1995, pp. 288-303; and Liu Jiaqi, "Penetration Technology for Tactical Missile Warheads," *Aerospace S&T Intelligence Studies Abstracts* (5), 1995, *CAMA*, Vol. 3, No. 6, 1996; Yang Bingwei, "Test Methods of Antirunway Penetrators," *Aerospace S&T Intelligence Studies Abstracts* (6), pp. 213-234. Yang Bingwei, from CALT's Beijing Institute of Special Electromachinery (*Beijing Teshu Jidian Yanjiusuo*), is the most prolific technical analyst on runway penetrators. The PLAAF is believed to have already fielded an anti-runway submunition cluster bomb.

62. See, for example, Li Xinyi, "On the Air Supremacy and Air Defense of Taiwan and China: Is Taiwan An 'Unsinkable Aircraft Carrier'?," *Taiwan de Junbei (Taiwan Military Preparations)*, July 1, 1996, pp. 11-18, in *FBIS-CHI*-97-323.

63. Gong Jinheng, "High Powered Microwave Weapons: A New Concept in Electronic Warfare," *Dianzi duikang jishu*, February 1995,

pp. 1-9. Gong is from the Southwest Institute of Electronic Equipment (SWIEE), China's premier electronic warfare research entity.

64. For a comprehensive overview of the technologies associated with HPM weapons, see Carlo Kopp, "The E-Bomb—A Weapon of Electrical Mass Destruction," in Winn Schwartau, *Information Warfare*, New York: Thunder's Mouth Press, 1994, pp. 296-297; also see J. Swegle and J. Benford, "State of the Art in High Power Microwaves: An Overview," paper presented at the Lasers 1993 International Conference on Lasers and Applications, Lake Tahoe, Nevada, December 6-10, 1993. Swegle and Benford point out that the United States, Russia, France, and the United Kingdom have HPM programs in addition to China. Zhu Youwen and Feng Yi, Gaojishu Tiaojianxia de Xinxizhan, (*Information Warfare Under High Technology Conditions*), Academy of Military Science Press, 1994, pp. 308-310; "Beam Energy Weaponry: Powerful as Thunder and Lightening," *Jiefangjun Bao*, December 25, 1995, in *FBIS-CHI*-96-039; "Outlook for 21st Century Information Warfare," *Guoji Hangkong*, (*International Aviation*), March 5, 1995, in *FBIS-CHI*-95-114; "Microwave Pulse Generation," *Qiang jiguang yu lizishu*, May 1994, in *JPRS-CST*-94-014. CAEP's Institute of Applied Electronics, University of Electronic Science and Technology of China, and the Northwest Institute of Nuclear Technology in Xian are three of the most important organizations engaged in the research, design, and testing of Chinese HPM devices. The PRC appears to have mastered at least two HPM power sources—the FCG and vircator. The greatest challenge is the weaponization process.

65. See Liu Shiquan, "A New Type of 'Soft Kill' Weapon: The Electromagnetic Pulse Warhead," *Hubei hangtian jishu* (*Hubei Space Technology*), May 1997, pp. 46-48. Liu is from the Sanjiang Space Industry.

66. Xu Licheng, "Research On Penetration Depth of Projectile into Thick Concrete Targets," *Qiangdu yu huanjing*, April 1996, pp. 1-7, *CAMA*, Vol. 3, No.1; and Zhu Bao. Xu is from the Beijing Institute of Special Electromechanics (14˙RI); for discussions on negating hardened targets, see Xu Xiaocheng, "Research on Penetration Depth of Projectiles Into Thick Concrete Targets," *Qiangdu yu huanjing*, 1996 (4), pp. 1-7. Also see Zhu Bao, pp. 9-19.

67. For references to FAE warheads, see *Lianhe zhanyi di erpaobing zuozhan*, p. 11; Yuen Lin, "The Taiwan Strait Is No Longer a Natural Barrier—PLA Strategies for Attacking Taiwan," *Kuang Chiao Ching*, April 16, 1996, in *FBIS-CHI*-96-104; and *Janes Strategic Weapons Systems 1998* section on the DF-15. Also see John Wilson Lewis and

Hua Di, "China's Ballistic Missile Programs: Technologies, Strategies, and Goals," *International Security*, Vol. 17, No. 2, Fall 1992, p. 33. An FAE is a variation of the napalm bomb that hits the ground, breaks open, and creates a mist of flammable liquid. A small delayed action explosive then goes off, causing the cloud to ignite. The pressure of the blast is sufficient to wreck aircraft, ships, and equipment as well as being fatal to personnel. The only other device to produce similar results is a nuclear weapon. Fortunately, FAE warheads are not as reliable as other types of bombs and weather conditions can seriously degrade their effects.

68. Zhu Yifan, Zhang Xuebin, and Wang Weiping, "ATBM Intercept Decision Modeling," *Guofang keji daxue xuebao*, January 1, 1999, pp. 29-32, in *FBIS-CHI*-1904, June 17, 1999.

69. Zhang Minde, "Simulation Research of Defenses Against Conventional Ballistic Missile Re-Entry Vehicles," *Xitong gongcheng yu dianzi jishu*, 1997, 19 (4), pp. 45-49. The simulation was conducted by CASC's Beijing Optoelectronic Engineering General Design Department. For general background on saturation, see Harshberger, pp. 169-170.

70. David Fulghum, "China Exploiting U.S. Patriot Secrets," *Aviation Week and Space Technology*, January 18, 1993, pp. 20-21.

71. Gan Chuxiong and Liu Jixiang, *Daodan yu yunzai huojian zongti sheji (General Design of Missiles and Launch Vehicles)*, Beijing: Defense Industry Press, January 1996, p. 42-43; and Wu Ganxiang, "Guowai fanjichang wuqi," (Foreign Antirunway Weapons), in Xu Dazhe, ed., *Guowai dandao daodan jishu yanjiu yu fazhan*, Astronautics Publishing House, 1998, pp. 65-76. The control maneuver may be necessary to slow down the reentry speed to allow acquisition of the target image in the ballistic missile's seeker.

72. Gan and Liu, p. 45. Also see Zhang Demin, "Study on Penetration Techniques on New Generation Ballistic Missiles," in *Xinjunshi gemingzhong daodan wuqi fazhan qianjing*, November 1996, pp. 18-24.

73. Gan and Liu, p. 45.

74. Wang Guobao, "Initial Discussion on Tactical Ballistic Missile Electronic Warfare," *Hangtian dianzi duikang*, April 1997, pp. 1-7 (*CAMA*).

75. Bill Gertz, *Betrayal*, Boston: Regnery Press, p. 254.

76. Li Qiang, "Current Status and Follow-On Development of Laser Cladding Wear-Resistance Coatings," *Yuhang cailiao gongyi*, January 1997, pp. 13-18. At least one institute involved in the testing is Harbin Institute of Technology. Also see Ji Shifan, "Laser Resistant Protection of Missiles," *Daodan yu hangtian yunzai jishu*, May 1996, pp. 35-42.

77. Jin Weixin, "Mathematical Modeling of Tactical Surface to Surface Missiles Against TMD," in *Systems Engineering and Electronic Technology*, 1995, 17 (3), pp. 63-68. Zhang Demin and Hou Shiming, ""Simulation Research of Offensive and Defensive Capability of Conventional Maneuvering Reentry Missile," *Xitong gongcheng yu dianzi jishu*, 1997, 19 (4), pp. 45-49. Full translation in *FBIS-CHI*-97-272. Zhang is from the Beijing Electromechanical Engineering Design Department, also known as the CASC Fourth Systems Design Department. According to one evaluation, PAC-2+ has a probability of kill of 10-25 percent against an unidentified tactical ballistic missile. See Zhao Yuping, "Probability of PAC-2 Intercepting a Certain Tactical Ballistic Missile," paper presented at November 1997 conference of National Missile Designers Specialist Network in *CAMA*, Vol. 5, No. 3.

78. Du Xiangwan, "Ballistic Missile Defense and Space Weapons," in *Quanguo gaojishu zhongdian tushu, jiguang jishu linghuo*, (*National High Technology Key Reference—Laser Technology Realm*).

79. *DoD Report to Congress on the Cross-Strait Security Situation*, February 1999.

80. For background on China's cruise missile development, including an organizational breakout of the Third Academy, see Mark A. Stokes, *China's Strategic Modernization: Implications for the United States*, Carlisle: Strategic Studies Institute, September 1999, pp. 79-86; and John Downing, "China Develops Cruise Missiles, *Jane's Defense Weekly*, February 1998.

81. Wang Jianmin and Zhang Zuocheng, "Jiasu jibenxing xiliehua jincheng nuli fazhan woguo feihang daodan shiye" (Rapid Progress in Series Development of China's Cruise Missile Industry), *Zhongguo hangtian*, September 1996, pp. 12-17. Some have estimated that a developing country like China could acquire at least 100 land attack cruise missiles at a cost of $50 million (i.e., $500,000 apiece). See Dennis M. Gormley and K. Scott McMahon, "Proliferation of Land-Attack Cruise Missiles: Prospects and Policy Implications," in Henry Sokolski, ed., *Fighting Proliferation: New Concerns for the Nineties*, Air University Press, 1996, pp. 131-167.

82. A land-attack Silkworm can readily fit within a 12-meter standard shipping container equipped with a small erector. See Dennis M. Gormley, "Cruise Missile Proliferation: Threat, Policy, and Defenses" Proliferation Roundtable, Carnegie Endowment for International Peace, October 9, 1998.

83. Two principles in particular are "basic serialization" (*jiben xingxiliehua*) that calls upon reliance on a basic airframe from which several upgraded variants can be derived. The second basic principle is "*sanbuqi*" that calls for having one system in production, one in applied R&D, and a generation-after-next in conceptual development. China's space and missile industry generally prefers to make incremental modifications to tried and trusted designs.

84. As a general rule, the Haiying series (HY-2, HY-3, HY-4) is surface-to-ship. The Yingji (YJ) designator is normally air-launched (i.e., YJ-6). There are exceptions—the YJ-8 can be ship or submarine launched (i.e., the YJ-82).

85. See "Zhongguo jingmi jixie jinchukou gongsi," (CPMIEC) *Xiandai junshi* (CONMILIT), 25th anniversary edition, 1996, pp. 16-23. The "XY" designation is likely a temporary one. Air-launched LACMs would be expected to have a "Yingji" designation. The initial Silkworm, the Haiying-2 (HY-2), utilized liquid propellant that limited its range to less than 100 kilometers. Variants incorporated radar or TV-guided terminal guidance systems. An air launched variant—the YJ-6 (C-601)—utilized the basic HY-2 airframe. The YJ-6 is launched from the B-6D bomber that has an operating radius of 1800-2000 kilometers. Like the HY-2, however, the YJ-6 range is less than 100 kilometers. In the 1980s, CASC's Third Academy developed an extended range Silkworm variant that integrated a turbojet engine (*wopen fadongji*). The turbojet variant has a range of up to 135 kilometers, is equipped with a 500-kilogram warhead, and can be launched from the B-6D or from shore. See Wang Jianmin, "Jiasu jibenxiliehua jincheng nuli fazhan woguo feihang daodan shiye," *Zhongguo hangtian*, September 1996, pp. 12-17.

86. Teal Group Corporation, "Chinese Anti-Ship Missiles," *World Missiles Briefing*, Fairfax, Virginia: Teal Group Corporation, May 1995, p. 2. One should not discount the possibility of extending the range through modest elongation of the fuselage that would provide space for addition fuel. There has been a recent flurry of reports of a different generation of LACM designated the CF-2000 or Hong Niao, a system based on the Russian Kh-55. See Duncan Lennox, "China's New Cruise Missile Programme 'Racing Ahead'," *Jane's Defence Weekly*, January 11, 2000.

87. Shirley Kan and Robert Shuey, "China: Ballistic and Cruise Missiles," *CRS Report for Congress*, March 21, 1997 (97-391), p. 11. For the Israeli connection, see "Israel to Equip Chinese Cruise Missile with Penetrator Warhead," *Flight International*, February 5-11, 1997, p. 13.

88. Undated China Precision Machinery Import and Export Corporation Marketing brochure, "C-201W Coast-to-Ship Missile System." The "W" probably stands for "wopen," or turbojet engine. According to Chinese aerospace publications, the U.S. Tomahawk Land Attack Missile (TLAM)-B utilizes a turbojet engine and has a range of 500 kilometers. The TLAM-C adopts a more efficient turbojet (*woshan*) engine that extends the range to 1200 kilometers. A turbofan engine (*woshan fadongji*) could significantly enhance the range of China's land attack cruise missiles. China's aviation industry has produced turbofan engines since the 1960s. The WS-7, a variant of the Rolls-Royce Spey Mk 202 engine, is used to power the H-7 fighter-bomber.

89. See Jason Glashow and Theresa Hitchens, "China Speeds Development of Missile With Taiwan Range," *Defense News*, March 4-10, 1996, p. 1; and Duncan Lennox, "China: Offensive Weapons," *Jane's Air Launched Weapons*, Surrey, United Kingdom: Jane's Information Group, 1996. By contrast, the U.S. Standoff Land Attack Munition (SLAM) is equipped with a 220kg warhead. Tomahawks have between a 320-480kg warhead.

90. Si Xicai, "Research on Long Range Antiradiation Missile Passive Radar Seeker Technology," in *Zhanshu daodan jishu* (Tactical Missile Technology), 1995, Vol. 2, pp. 42-52; other studies on specific approaches to ARM technology include Yang Huayuan, "Study on Superwideband High Accuracy Microwave DF System," in *Daojian yu zhidao xuebao*, February 1995, pp. 7-12. There are also strong indications that SAST's system engineering organization, the Shanghai Institute of Electro-Mechanical Engineering, is carrying out preliminary R&D on a long-range air-to-air anti-radiation missile for targeting airborne early warning platforms, such as the U.S. E-3 AWACS or Taiwan's E-2Ts. Engineers note critical technologies for development of a long-range ARM include a passive seeker with a sensitivity of greater than 100dB, as well as monolithic microwave (*danpian weibo*), gallium arsenide, very large scale, and very high speed integrated circuits (MMIC, GAAS, VLSIC, VHSICs). The seeker makes up greater than 50 percent of the R&D and production costs for an ARM. At least one Second Academy entity that has conducted work on antiradiation missile seeker technology is the Beijing Institute of Remote Sensing Equipment (probably the CASC 25th Research Institute).

91. *Jane's Strategic Weapons Systems*, 1998, People's Republic of China; Jane's Air-To-Ground Missile Programs. There are several systems that CASC appears to be studying to serve as the basis for an indigenous version: The U.S. AGM-88 HARM utilizes a solid motor and has a range of 40 kilometers. The UK's ALARM has a range of 70 kilometers; Israeli's STAR-1 uses a small turbojet engine, has a range of 100 kilometers and weighs only 182 kilograms. Chinese engineers note the Kh-31P has both long-range (i.e., 200km) air-to-air and air-to-ground variants.

92. Lu Xiaohong, "Launch Technology for Air-launched Antiradiation Missiles," *Astronautics Information Research*, HQ-97038, Astronautics Publishing House, December 1997. Lu is from Third Academy's Beijing Institute of Special Machinery, which is responsible cruise missile launching technology.

93. For one reference on land attack cruise missile and supporting GIS efforts, see Xu Haijiang, "Cruise Missile Mission Planning Research," *Astronautics Information Research*, HQ-97020, in *CAMA*, Vol. 5, No.1, December 1997. The author is from the Beijing Institute for Electromechanical Engineering.

94. A GPS receiver is available in the United States for as little as $5,000; and radar altimeter for 2,500; and an IMU for $20-30,000. A flight management computer could involve a miniaturized $2,500 commercial system with software to permit flight control, autopilot functions, onboard system monitoring, and flight path and course navigation.

95. This process is described in Edward Harshberger, *Long Range Conventional Missiles: Issues for Near Term Development*, Santa Monica: RAND, 1991, pp. 46-50. Also see Zhou Rui, "Image Guidance Aimpoint Selection System," *Zhanshu daodan jishu*, January 1996, pp. 32-36.

96. Zhong Longyi, "Zuhe daohang xitong he bingxing duoji xitong zai xunhang daodanzhong de yingyong," (Application of Combined Navigation Systems on Cruise Missiles) in *Hangtian qingbao yanjiu (China Information Research)*, 1993 (3), pp. 432-445. Zhong is from the Third Academy's 8357 Research Institute, responsible for cruise missile control systems. One of China's first research projects on digital image processing, written by Sun Zhongkang and Shen Zhenkang, was published by the PLA National Defense University in 1985.

97. Guan Dexin, "The Investigation of Compatible Receiver For GPS and GLONASS," *Xitong gongcheng yu dianzi jishu*, 1996, Vol. 18, No. 7,

pp. 69-74; and Sheng Jie, "Demonstration of Navigation Performance of GLONASS/GPS Composite Receivers," *Weixing yingyong*, February 1994, pp. 56-59.

98. Zheng Wanqian, p. 43; The PLA likely has been developing an imagery library that could support DSMAC targeting for several years. TERCOM requires highly sophisticated digital mapping systems and powerful computers. COSTIND and the Second Artillery have made significant achievements in both areas. See Wang Yongming, "Introduction to Military Electronic Maps," *Xiaoxing weixing jisuanji xitong*, (*Mini-Micro Computer Systems*), August 1995, pp. 12-18, in *FBIS-CST*-96-001. Wuhan Technical University of Surveying and Mapping is one institute involved in digital mapping. Also see Jing Shaoguang, "GPS/SINS Integrated Navigation System for Cruise Missiles," *Xibei gongye daxue xuebao*, 1997, 15 (1), pp. 79-83. At least one State Laboratory is dedicated to R&D on scene matching technology—the Image Information Processing and Intelligence Control Laboratory, Imaging Institute, under Huazhong University in Wuhan.

99. One Taiwan source explicitly asserts land attack cruise missiles will be assigned to the Second Artillery. See "Mainland Acquisition of Russian Weapons Viewed," *Lien-ho pao*, April 29, 1996, in *FBIS-CHI*-96-086. In support of this new mission, the Second Artillery's Fourth Research Institute has been modeling the ability of cruise missiles to penetrate air defense systems. See Sun Xiangdong and Qin Xiaobo, "Operational Efficiency Analysis of Cruise Missiles Against SAMs," in *Xitong gongcheng yu dianzi jishu* (*Systems Engineering and Electronics*), October 1996, pp. 59-63, in *FBIS-CST*-97-013. Sun and Qin are from the Second Artillery's Fourth Research Institute.

100. Liu Kejun, "Information Warfare Challenge Faced By Navy," *Zhongguo dianzi bao*, October 24, 1997, p. 8, in *FBIS-CHI*-98-012.

101. Tian Baolong and Li Wengang, "Feihang daodan CAM chejian danyuan xitong" (Cruise Missile CAM Workshop Unit System), *Zhongguo hangtian*, April 1993, pp. 44-46; Xu Haijiang, "Virtual Reality and Its Application in Development of Cruise Missiles," in *Feihang daodan*, 1996 (8), pp. 1-9; Wang Zhenhua, "Parallel Computation on Supercomputers for Axisymmetric Interaction Flow," *Yuhang xuebao* (*Journal of Astronautics*), January 1995, pp. 43-45, in *JPRS-CST*-95-005.

102. Li Weiliang, "Jiang Zemin dao Beijing fangzhen zhongxin zhouyan" (Jiang Zemin Inspects Beijing Simulation Center), *Zhongguo*

hangtian bao, January 17, 1994, p.1; Li Li, "Chinese Simulation Technology Among Leaders Worldwide," *Liaowang zhoukan*, August 16, 1993, pp. 4-5, in *JPRS-CST*-93-017. American aerospace representatives who have been allowed access have remarked that the CASC Beijing Simulation Center is very close in capabilities to Western simulation facilities.

103. *London Quds Press*, February 9, 1999, in *FBIS-CHI*-1441-99.

104. Stokes, p. 82-83.

105. If the Chinese are looking to develop a 1,500-kilometer missile, the Russian 1,500-kilometer range AS-15 could be used as a model. Some modifications would have to be made to enable it to launch from the ground. The Tomahawk has a 450-kilometer range, while the U.S. AGM-86B has a 3,000-kilometer range.

106. Russia's Raduga Design Bureau has reportedly assisted the Third Academy in application of stealth technology to an unidentified air launched cruise missile. See "Russian Missile Assistance to China," *Flight International*, August 31, 1995.

107. In other words, radar that can pick up an airborne target at 200 kilometers will now only be able to detect the target at 50 kilometers, resulting in less reaction time. Undated brochure, "Xikai (Zhongguo) guangxue jishu youxian gongsi," (See China Optical Technology Company). The brochure notes that the radar absorbing material, designated BD-21/SF-18, can reduce a target such as a cruise missile to an RCS of 0.1 square meters (-10 dB). The absorbing material and structural modifications can reduce the RCS to -30 dB.

108. Zhang Haixiong, "ADN: Oxidizer for A Low Signature Propellant," in *Feihang daodan*, July 1996, pp. 35-38, in *CAMA*, 1996, Vol. 3, No. 6; and Lu Xiaohong, "Camouflage and Concealment Technology of Mobile Missile Launchers and Ground Equipment," in *Harbin Institute of Technology Journal*, December 1996, pp. 266-277. Lu is from the Third Academy's Beijing Institute of Special Machinery, responsible for cruise missile launchers.

109. See Wang Jianmin, "Work Hard to Develop Cruise Missile Industry," *Zhongguo hangtian*, September 1996, pp. 12-17; Sun Qingguang, "Study on Laser Imaging Guidance," *Feihang daodan*, March 1995, pp. 46-50, in *CAMA*, 1995, Vol. 2, No. 3; Liu Yongchang, "Infrared Imaging Precision Seeker Technology," *Hongwai yu jiguang jishu* (Infrared and Laser Technology), 1996, Vol. 25, No. 3, pp. 47-53, in *CAMA*, Vol. 3, No. 6; Zhao Jun, "Applied Research Into Laser Imaging

Guidance Technology Development," *Hangtian qingbao yanjiu*, HQ-96039, in *CAMA*, 1997, Vol. 4, No. 2; and Li Jin, Development of Infrared Focal Plane Array Imaging Technology," in *Feihang daodan qingbao yanjiu baogao wenzhai (Cruise Missile Information Research Reports)*, December 1996, pp. 190-209, in *CAMA*, Vol. 4, No. 6. Leading the infrared imaging effort is the Third Academy's Tianjin Jinhang Technical Physics Institute.

110. Sun Qingguang, "Jiguang chengxiang zhidao ji ganrao moushi de yanjiu" (Research Into Laser Imaging Guidance and Jamming), *Hangtian qingbao yanjiu*, HQ-93017, pp. 228-241.

111. Zheng Wanqian, p. 43.

112. Bases are located at Shenyang (80301 Unit); Huangshan (80302 Unit); Kunming (80303 Unit); Luoyang (80304 Unit); Huaihua (80305 Unit); and Xining (80306 Unit). The Second Artillery has one engineering design academy and four research institutes to solve problems associated with operations, TELs, and logistics (First Institute), command automation, targeting, and mapping (Third Institute), and missile and warhead engineering design (Academy of Engineering Design). The Second Artillery's Command College in Wuhan prepares officers for leadership positions within headquarters elements and launch brigades. The Engineering College in Xian educates technicians associated with equipment and technology departments at various headquarters and field units. General Second Artillery organizational information is drawn from numerous sources, to include open and internal (*junnei*) Chinese publications and from discussions while assigned as the assistant air attaché in Beijing, China from 1992-1995. Also see *PLA Directory of Personalities*, USDLO Hong Kong, 1996, pp. 48-51; Bill Gertz, "New Chinese Missiles Target All of East Asia," *Washington Times*, July 10, 1997, p. 1; Hisashi Fujii, "Facts Concerning China's Nuclear Forces," *Gunji Kenkyu*, November 1995, in *FBIS-CHI*-96-036; "Guangrong bang" (Outstanding Units), *Flying Eagle (Changying)*, November 3, 1993; "Guangrong bang" (Outstanding Units), *Flying Eagle*, May 1992; Lewis and Xue, p. 213, footnote; and Nuclear Weapons Databook, Vol. 5, pp. 324-335. Among sources, *Flying Eagle*, one of a handful of Second Artillery-associated publications, is most useful in piecing together the organizational structure. Second Artillery organizational issues are also discussed in author's *Strategic Modernization* monograph.

113. "The Strategic Nuclear Force Organization," in *Guojia junzhixue (The Science of the State Military System)*, undated, p. 3.

114. Stokes, pp. 94-95. The 80302 Unit's first conventional DF-15 SRBM brigade is said to be garrisoned in Leping, Jiangxi province. Taiwan sources indicate that the first extended range DF-11 brigade is being formed in the Yong'an/Nanping area. See "Military Officials Discuss PRC's Fujian Missile Positions," *Chung-yang jih-pao*, December 1, 1999, p. 10, in *FBIS-CHI*-0322, December 1, 1999. Bill Gertz from the *Washington Times* also reports that the second Second Artillery brigade in being formed near the city of Yong'an, Jiangxi province. A third brigade consisting of the DF-11/CSS-7 is under construction in Fujian province near the town of Xianyou. See Bill Gertz, "China Targets Taiwan with 2nd Missile Base, " *Washington Times*, December 8, 1999, p. 1. According to an unsubstantiated *Washington Times* article, the 80302 Unit is replacing its older DF-3 missiles with the DF-21. Whether or not these DF-21s will eventually have a conventional mission is unknown. See Bill Gertz, "New Chinese Missiles Target All of East Asia," *Washington Times*, July 10, 1997, p. 1.

115. *Lianhe zhanyi di erpaobing zuozhan*, p. 4. Another article supports the assertion that conventional Second Artillery units would be subsumed into the theater command structure, but notes that Beijing may direct operations though the Second Artillery chain-of-command. See Li Junsheng, "Lianhe zhanyi didi changgui daodan budui zuozhan zhihui wenti tantao" (Inquiry Into Joint Conventional Theater Surface-to-Surface Missile Unit Operational Command Problems), in *Lianhe zhanyi yu junbingzhong zuozhan (Joint Theater and Service Operations)* Beijing: National Defense University Press, 1998, pp. 228-231. Li is from an unidentified (probably Second Artillery) Third Research Institute.

116. *Ibid*, p. 5. During peacetime, these units are subordinate to the base headquarters.

117. *Ibid*, p. 4. The equipment assurance sub-units, the transfer point, and the transport may be the responsibility of a battalion-level "technical unit" *(jishu ying)*. A nuclear brigade's technical battalion manages a warhead station *(dantizhan)*, an inspection station *(zhuangjianzhan)*, and a technical service station *(jishu qinwuzhan)*. See "Guangrong bang," *Flying Eagle*, undated (probably 1993), p. 11.

118. For reference to a fourth battalion within a Second Artillery brigade structure, see "Guangrong bang" (Glorious Honor Roll), *Flying Eagle*, November 2, 1993, p. 10.

119. Senior Colonel Wang Benzhi, "Didi changui daodan huoli yunyong de jige wenti," (Some Questions Related to the Use of Conventional Surface-to-Surface Missile Firepower), in *Lianhe zhanyi*

yu junbingzhong zuozhan, (Joint Theater and Service Operations), Beijing: National Defense University Press, 1998, pp. 236-241. Colonel Wang is the Chief of Staff of the Second Artillery Huaihua Base (80305 Unit). One source states that an operational zone could be 20-40 square kilometers. It is unclear what echelon would operate in this size zone. See Lu Xiaohong, "Daodan jidong fashe zhuangbei ji dimian shebei weizhuang yu yinshen jishu Fenxi" (Analysis of Mobile Missile Launch and Ground Equipment Camouflage and Stealth Technology), in Xu Dazhe, *Guowai dandao daodan jishu yanjiu yu fazhan (Study and Development of Foreign Ballistic Missile Technology)*, Beijing: Astronautics Press, October 1998, pp. 193-202.

120. Mao Guanghong, "On Electromagnetic Management of the Modern Battlefield," *Jiefangjun bao*, May 21, 1996, p. 6, in *FBIS-CHI*-96-134.

121. Zhang Jian, "Analysis of ECCM Principles of Spread Spectrum Unified Satellite Tracking, Telemetry, and Control Network," *Hangtian dianzi duikang*, April 1997, pp. 26-30. Zhang is from the China Academy of Engineering Physics' Electronic Engineering Institute. Also see Wei Chenxi, "ECCM Measures for Military Communications Satellites," in *Hangtian dianzi duikang*, March 1997, pp. 31-34.

122. Qin Zhongping and Zhang Huanguo, "ALT: Algorithim for Attacking Cryptosystems," *Jisuanji Xuebao*, Vol. 20, No. 6, pp. 546-550, in *FBIS-CHI*-97-311; and Zhou Hong and Ling Xieting, "Encryption By Inverse Chaotic Systems," *Fudan xuebao*, June 1997, Vol. 36, No. 3, pp. 301-308, in *FBIS-CHI*-97-281.

123. Gan and Liu, pp. 42-45.

124. Lu Xiaohong, "Camouflage and Concealment Technology for Mobile Launchers and Ground Equipment of Strategic and Tactical Missiles," Aerospace Industry Press, HQ-96034, 1996. The key institute for CCD technology related to missile launchers is the Beijing Institute of Special Machinery. Wen Longzhi, "Evaluation of the Strategic Missile Survivability," in *Aerospace Science Intelligence Studies Report Abstracts*, No. 5, 1995, pp. 353-368.

125. Li Chunshan, "Introduction and Explanation of the National Military Standard 'Camouflage Requirements for Surface-to-Surface Missile Weapon Systems'," in *Hangtian biaozhunhua* (Space Standardization), 1994, Vol. 5, pp. 12-15. Li is from the Beijing Space Systems Engineering Design Department.

126. Kang Qing, "IR Stealth of Buried Targets," *Hongwai jishu*, 1996, 18 (6), pp. 21-24. Kang is from the PLA Academy of Logistics Engineering.

127. R&D of synthetic aperture radar satellite jammers is the specialty of Southwest Institute of Electronic Equipment (SWIEE). See Chen Ning, "Jamming Technology Against Synthetic Aperture Radar Satellites," *Hangtian Dianzi Duikang*, 1997 (4), pp. 45-48.

128. Chou Kuan-wu, "China's Reconnaissance Satellites," *Kuang chiao ching*, March 16, 1998, pp. 36-40, in *FBIS-CHI*-98-098. *Kuang chiao ching*, or *Wide Angle*, is a Hong Kong-based publication with close ties to the PRC military establishment. Official U.S. Government reports are consistent with this assessment. The 1998 Report to Congress on PRC Military Capabilities (pursuant to Section 1226 of the FY98 National Defense Authorization Act) states China already may possess the capability to damage, under specific conditions, optical sensors on satellites that are very vulnerable to damage by lasers. However, given China's current interest in laser technology, it is reasonable to assume that Beijing would develop a weapon that could destroy satellites in the future.

129. Wu Jinliang, "Range Testing of Satellite Communication Countermeasures," in *Electronic Countermeasure Technology and Intelligence Reform Abstracts*, November 1995, pp. 96-101. Reference to a Chinese study on GPS jammer is included in author's unpublished report, *China's Space and Missile Industry*, June 1995.

130. Stokes, pp. 118-119. One should note that in the 1980s, the United States considered modification of the Pershing-2 for ASAT missions, a system similar to the DF-21.

131. Xu Hui and Sun Zhongkang, "Temperature Differences Between Satellites and Satellite Decoys," *NUDT Journal*, 1994, Vol. 16, no. 3; also see Li Hong, "Identification of Satellites and Its Decoys Using Multisensor Data Fusion," *Xiandai fangyu jishu*, June 1997, pp. 31-36. Li is from the NUDT Electronic Technology Department.

132. *DoD Report to Congress on the Cross-Strait Security Situation*, February 1999.

133. Wang Chunyuan, *China's Space Industry and Its Strategy of International Cooperation*, Stanford University, Center for International Security and Arms Control, July 1996, p. 4; "China Building Satellite Tracking Station on Tarawa," *Asian Defense Journal*,

March 1997, p. 66; and "Satellite Command Station Operational in Kiribati," *Zhongguo xinwenshe*, October 14, 1997, in *FBIS-CHI*-97-287.

134. Trip report, NASA visit to China, June 12-22, 1991. For example, China plans to develop a 500-meter aperture radio space telescope for deep space exploration. With a price of approximately 25 million dollars, the system, which will be based in Guizhou Province, will primarily support civilian academic research, but could also be used to supplement China's space surveillance network. CAST and the China Academy of Sciences are involved. See "Beijing Plans to Develop 500 Meter Radio Telescope," *Xinhua,* April 9, 1998, in *FBIS-CHI*-98-099.

135. A 1993 edition of the Second Artillery journal *Flying Eagle* discussed a "comprehensive satellite early warning information management system" (*erpaobing weixing linkong yubao zonghe xiaoxi chuli xitong*) that began operations in as early as 1991.

136. *Lianhe zhanyi di erpaobing zuozhan*, p. 17.

137. *Ibid*, p. 10; and Guan Lin'gen, "Brief Analysis of Combined Fire Assault," *Jiefangjun Bao*, April 21, 1998, p. 6, in *FBIS-CHI*-0519-98. Some Western observers have asserted the initial phase would include strikes against the general population and infrastructural targets, such as power plants, fuel, industry, and transportation hubs as a means to weaken overall national resolve. However, the effects from these targets would take awhile to materialize. The PRC objective would be to achieve military dominance over Taiwan within two weeks to a month, before negative international economic and political developments can occur. For comments on the importance of strikes against enemy intelligence and electronic attack facilities in support of information dominance, see Yang Zhiguo, "Didi changgui daodan budui zhanfa chutan" (Initial Discussion of Surface-to-Surface Missile Unit Doctrine), in *Lianhe zhanyi yu junbingzhong zuozhan*, (*Joint Theater and Service Operations*), Beijing: National Defense University Press, 1998, pp. 242-245. Senior Colonel Yang is Chief of Staff of the Second Artillery's Luoyang Base (80304 Unit).

138. The PLAAF appears to be placing more emphasis on developing a deep strike capability. In 1995, the PLAAF conducted a major strike exercise in the Lanzhou Military Region. The exercise involved a Red force strike package that conducted a night mission from a distance of 1,000 kilometers to strike the Blue force's air base. In October 1995, a conference, chaired by GSD DCS LTG Wu Quanxu was held at LMR HQ to review the exercise and associated doctrinal development issues. A more complex strike exercise was carried out in northwest China in

September 1996 (Exercise 96-9) when the PLAAF used multiple types of aircraft (i.e. A-5s, B-6s, F-7s, F-8IIs, and SU-27s) organized into composite formations. The strike package included electronic countermeasures, strikes against enemy missiles, airfields, and radar. This is an initial indication that the PLAAF could be shifting from an exclusively air defense mission to one including long-range strike missions. Like the USAF, the PLA views offensive counterair missions as an integral aspect of air defense. Lanzhou MR exercise areas appear to serve as the primary test bed for evolving doctrinal development. See Mei Lin, "PLA Methods of Operations Assessed," and Zhang Lianfu, "'96-9' Yanxi," *Zhongguo kongjun*, May 1998.

139. Senior Colonel Wang Benzhi, pp. 236-241.

140. *Lianhe zhanyi di erpaobing zuozhan*, p. 10. One source indicates that during annual meetings at Beidaihe in August 1999, China's senior leadership decided to accelerate the production and deployment of enough ballistic missiles to outfit four SRBM brigades by 2002. As of January 2000, four batches of DF-15 SRBMs had been completed. Each production batch (*piliang shengchan*) consists of 60 missiles, indicating that 240 DF-15 missiles have been produced. See Will Young, "Shenmi de zhongguo daodan budui," (The Development of the Chinese Second Artillery), *Shijie junshi luntan* (*World Military Forum*), Internet edition in Chinese (http//:www.wforum.com), January 2000. The rapid growth in the Second Artillery's conventional force has prompted the outsourcing of training. In 1999, the Northwest China Engineering University and the Second Artillery agreed to jointly foster more high quality personnel for China's missile forces.

141. Guan Lingen, "Brief Analysis of Combined Fire Assault," *Jiefangjun bao*, April 21, 1998, p. 6, in *FBIS-CHI*-0519-98. In comparison, allied forces in the Gulf War used 137 theater missiles (TLAMs/CALCMs) during the first 24 hours of the conflict. Each wave consisted of around 50 missiles. Western reporting indicates the PLA currently has only one brigade consisting of 150-200 SRBMs. See Tony Walker and Stephen Fidler, "China Builds Up Missile Threat," *Financial Times*, February 10, 1999, p. 1.

142. Guan Lingen, "Brief Analysis of Combined Fire Assault," *Jiefangjun bao*, April 21, 1998, p. 6, in *FBIS-CHI*-0519-98.

143. See Sun Xiaohe, "Jiaqiang huoli xietiao, fahui zhengti weili" (Strengthen Firepower Coordination, Give Play to Comprehensive Power), in *Lianhe zhanyi yu junbingzhong zuozhan*, (*Joint Theater and Service Operations*), Beijing: National Defense University Press, 1998,

pp. 281-285. Senior Colonel Sun is Deputy Director of the Guangzhou Military Region Service Arms Department.

144. *Lianhe zhanyi di erpaobing zuozhan*, p. 10. Also see Wang Xuejin and Zhang Huaibi, "Didi changgui daodan budui zuozhan zhidao sixiang fenxi," (Analysis of Conventional Surface-to-Surface Missile Operations Guiding Thought) in *Lianhe zhanyi yu junbingzhong zuozhan, (Joint Theater and Service Operations)*, Beijing: National Defense University Press, 1998, pp. 223-227. Wang and Zhang call the strike phase the "operations implementation phase" (*zuozhan shishi jieduan*).

145. Most of the critical targets in a Taiwan scenario are static and would not change significantly over time. Therefore, a satellite revisit rate of a few days, or even weeks, could be sufficient. China's commercially acquired imagery could meet its strategic targeting requirements. However, with the possible exception of Russia, Beijing could not rely on foreign sources of imagery after initiation of hostilities. For follow-on tactical strike missions, domestic imaging satellites would be needed for theater reconnaissance and warning. The projected five-meter spatial resolution of China's EO/SAR satellite constellation would support most PLA targeting requirements.

146. Open sources indicate the DF-15s are likely transported to assembly areas to via rail. While the DF-15 TELs are road mobile, the DF-15 MAZ-543-like TEL is limited to a maximum of 63 kilometers an hour on an open highway. Barring major infrastructure investments, road conditions and traffic in this area, however, are not ideal for rapid and distant deployment of 20-ton TELs and a quiver of 3-ton missiles. In addition, road travel significantly increases the chances of detection. There is a major 15-year project underway to expand the rail network in Fujian and Jiangxi provinces that will increase the number of available launch sites and complicate the tracking of the missiles on the ground. Highest priority is being given to linking Nanping to Hengfeng/Shangrao, creating a racetrack bounded by Yingtan, Shaowu, Nanping, Shangrao/Hengtian, and back to Yingtan. Funding in part is being provided by Japan. See "Fujian Seeks Foreign Funds for Railroad Construction," *Xinhua*, February 12, 1996, in *FBIS-CHI*-96-029; and "Fujian Governor Announces Plans For Six New Railways," *Xinhua*, August 1, 1997, in *FBIS-CHI*-97-213. For comments on Leping garrison, Yong'an launch site, and use of rails, see "Defense Ministry Analyzes 4th Missile Launch," *China Broadcasting Corporation News Network*, March 13, 1996, in *FBIS-CHI*-96-051.

147. *Lianhe zhanyi di erpaobing zuozhan*, p. 17. Theater command authorities would determine a deployment pattern that would be

centered on the brigade's mobile command center. Also see Richard D. Fisher, "China's Missiles Over the Taiwan Strait: A Political and Military Assessment," paper presented at September 96 Coolfont Conference on the PLA, pp. 1-30. For reference to a unit having an assigned operating area, see Senior Colonel Wang Benzhi, pp. 236-241.

148. Zhu Bao, pp. 9-19.

149. Zhang Hu, "Application of GPS in Missile Maneuvering Positioning," *Zhongguo yuhang xuehui fashe gongcheng yu dimian shebei wenzhai (China Astronautics Society Launch Engineering and Ground Equipment Abstracts)*, November 1993.

150. Ge Xinqing, Mao Guanghong, and Yu Bo, "Xinxizhan zhong daodan budui mianlin de wenti yu duice," (Questions and Countermeasures Facing Missile Units in Information Warfare), in *Wojun xinxizhan wenti yanjiu (Studies on Chinese Information Warfare Issues)*, Beijing: National Defense University, pp. 189-192. The authors are from the Second Artillery's Command Academy in Wuhan. It should be noted that Fujian province by itself has 16,000 kilometers of fiber optic cable.

151. Wang Jixiang and Chang Lan, "Guowai jidong dandao daodan dimian shengcun nengli yanjiu," (Study on Survivability of Foreign Mobile Ballistic Missiles), in Xu Dazhe, *Guowai dandao daodan jishu yanjiu yu fazhan (Study and Development of Foreign Ballistic Missile Technology)*, Beijing: Astronautics Press, October 1998, pp. 96-108. Wang and Chang are from CALT's systems integration department. The article describes foreign capabilities but concludes with specific recommendations for China. Chinese defense industries have developed a range of tactical communications systems, including mobile 1-3 meter very small aperture terminal (VSAT) satellite communication dishes and highly directional tactical digital microwave system. VSAT dishes are optimized for Ku- or L-Band satellite communications. Based on author's discussions in Beijing with foreign diplomats in 1995, the Second Artillery has been particularly interested in steerable spot beam satellites. According to its brochure, the tactical microwave system, produced by Shenyang Huitong Electronic Research Institute, has a 50-kilometer range. At the end of 1997, culminating a 3-year effort, the Second Artillery's Communications Department completed the acceptance testing of a new digital microwave communications system. VSAT systems are produced by a wide range of manufacturers. One tactical VSAT system, outlined in another brochure, is a mobile 3-meter dish produced by the Nanjing Research Institute of Electronics Technology. Use of digital microwave at the company level would indicate that launchers could be limited to an operating area of within

50 kilometers of the battalion command center. It is not clear, however, if such a communications mode would be assigned to such a low echelon. For reference to the automated C2 system, see Han Tiejun and Li Qinsuo, "Didi changui daodan budui zuozhan de jiben yuance," (Fundamental Principles of Conventional Surface-to-Surface Missile Unit Operations), in *Lianhe zhanyi yu junbingzhong zuozhan*, (*Joint Theater and Service Operations*) Beijing: National Defense University Press, 1998, pp. 232-235.

152. See Wang Jixiang and Chang Lan, pp. 96-108. Pre-Gulf War estimates assessed that it would take approximately one half hour to move a transporter erector launcher after it launched its missile. The reality was that the Iraqis were able to do this in 4 to 5 minutes.

153. A brigade consists of at least four battalions probably with three to four companies. Each company likely is responsible for at least one launcher. If one assumes a notional structure of four battalions per brigade with four companies/launchers each—then a brigade would be able to execute a raid size of at least 16 SRBMs at one time. Seven Second Artillery brigades, equipped with a mix of SRBMs, LACMs, and MRBMs, could notionally achieve a raid size of at least 112 theater missiles. Three salvos would utilize 336 missiles. Remaining theater missiles in the PLA arsenal would likely be kept in reserve for other contingencies and/or to support naval operations and amphibious landings. The Chief of Staff of the 80305 Unit in Huaihua, Hunan province refers to only two salvos in the opening phases of a conflict (see next footnote). See Senior Colonel Wang Benzhi, pp. 236-241.

154. *Lianhe zhanyi di erpaobing zuozhan*, p. 17. The concept of synchronized, multi-axis strikes is a fundamental principle of Second Artillery conventional doctrine (*duodian, duofangxiang, tongshi tuji*). Other important operational concepts discussed by Senior Colonel Wang from Huaihua include "*xushi bingyong, shengdong xiji,*" (literally "use reality, make a noise in the east, but strike to the west"); and "*xiaojiange, duoboci tuji*" (literally "cut time and strike in multiple waves"). The first calls for integration of simultaneous launches from different launch azimuths and use of infrared radiation "disruption" to confuse enemy satellite early warning systems and complicate enemy attack operations. The second includes use of two strike waves, the first "screening" the second by exploiting "time lags" (*shijiancha*) in missile defenses. See Senior Col Wang's "Didi changgui daodan huoli yunyong de jige wenti."

155. *Ibid.*

156. "The U.S. Military's Three Choices on Intervention," in *Taiwan de Junbei*, July 1, 1996, pp. 76-79, in *FBIS-CHI*-97-302.

157. See Li Xinyi, "On the Air Supremacy and Air Defense of Taiwan and China: Is Taiwan An 'Unsinkable Aircraft Carrier'?," *Taiwan de Junbei (Taiwan Military Preparations)*, July 1, 1996, pp. 11-18, in *FBIS-CHI*-97-323. For a general discussion on the combined use of ballistic and land attack cruise missiles, electronic warfare, and air strikes, see Yuan Lin, "The Taiwan Strait is No Longer A Barrier—PLA Strategies for Attacking Taiwan," *Kuang chiao ching (Wide Angle)*, April 16, 1996, No. 283, pp. 14-19. *Wide Angle* is a Hong Kong-based publication with close association with the PLA.

158. The PRC has closely studied the effectiveness of missile defense systems and is investing in developing the capacity to jam Taiwan's PATRIOT-like Modified Air Defense System radar. See for example, Xiao Shunping, Wang Guoyu, and Ma Jianwu, "Detection Simulation Modeling for PATRIOT Radar Networks," *Guofang keji daxue xuebao (Journal of the National University of Defense Technology)*, January 1, 1999, pp. 33-36, in *FBIS-CHI*-1924, June 17, 1999.

159. Estimated missile requirements are drawn from Edward R. Harshberger, *Long Range Conventional Missiles: Issues for Near-Term Development*, RAND: Santa Monica, 1991, p. 183.

160. Senior Colonel Wang Benzhi, "Didi changui daodan huoli yunyong de jige wenti," (Some Questions Related to the Use of Conventional Surface-to-Surface Missile Firepower), in *Lianhe zhanyi yu junbingzhong zuozhan, (Joint Theater and Service Operations)*, Beijing: National Defense University Press, 1998, pp. 236-241. Colonel Wang is the Chief of Staff of the Huaihua Base (80305 Unit) in Hunan Province.

161. The PLA could attempt to limit the useable segment of a runway to less than the minimum takeoff distance for the aircraft using them. Generally, for example, an attack on a standard 10,000-foot runway would attempt to cut the runway into three segments, permitting only 3,000-3,500 feet per segment. One source asserts that it would take 15-48 missiles to close a base, assuming a 50-meter CEP. An increase in CEP would reduce the number of required missiles. See David Blair, "How To Defeat the United States: The Operational Military Effects of the Proliferation of Weapons of Precise Destruction," in Henry Sokolski, *Fighting Proliferation: New Concerns for the Nineties*, Air University Press, pp. 75-94. Also see Senior Colonel Wang Benzhi, "Didi changui daodan huoli yunyong de jige wenti," (Some Questions Related to the Use of Conventional Surface-to-Surface

Missile Firepower), in *Lianhe zhanyi yu junbingzhong zuozhan*, (*Joint Theater and Service Operations*), Beijing: National Defense University Press, 1998, pp. 236-241.

162. For a superb discussion of airbase attacks, see Christopher M. Center, "Ignorance is Risk: The Big Lesson From Desert Storm Air Base Attacks," in *Airpower Journal*, Winter 1992, pp. 25-35; and Captain Peter C. Bahm and Captain Kenneth W. Polasek, "Tactical Aircraft and Airfield Recovery, *Airpower Journal*, Summer 1991, pp. 32-45. It is worthy to note that during the October 1973 Arab-Israeli War, Arab repair teams restored runways in just 9-12 hours. The Iraqis were able to repair runways in as little as 4 to 6 hours.

163. See Wang Jixiang and Chang Lan, p. 107.

164. Lu Ting-hua, "Simulated Attack on Taiwan?," *Tsu-li Wan-pao*, April 28, 1999, p. 1, in *FBIS-CHI*-639-99, April 29, 1999.

165. Harshberger, p. 183.

166. Xu Minfei, Zhu Zili, and Li Yong, "Feasibility of Technologies for Use of Ballistic Missiles to Counter Aircraft Carriers," *Guofang Keji Cankao*, 1997, 18(4), pp. 126-130, summarized in *CAMA*.

167. Wang Guobao, "Initial Discussion on Tactical Ballistic Missile Electronic Warfare," *Hangtian dianzi duikang*, April 1997, pp. 1-7, summarized in *CAMA*.

168. See Wang Jixiang and Chang Lan, p. 107. Most vulnerable would be Kadena AB and Yokosuka Naval Base in Japan.

169. For a good summary of Second Artillery CCD practices, see Ge, *et al.*, "Xinxizhan zhong daodan budui mianlin de wenti yu duice," pp. 189-192. Also see Wang Jixiang and Chang Lan, pp. 96-108. Wang and Chang are from the Beijing Institute of Astronautical Systems Engineering. See also Lu Xiaohong, "Camouflage and Concealment Technology for Mobile Launchers and Ground Equipment of Strategic and Tactical Missiles," Aerospace Industry Press, HQ-96034, 1996. The key institute for camouflage, concealment, and deception (CCD) technology related to missile launchers is the Beijing Institute of Special Machinery (CALT 15th Research Institute).

170. The system is probably known as the "Bodyguard," unveiled at IDEX '97 Arms Show. The brochure notes that the Bodyguard is a mobile system consisting of four vehicles. The Bodyguard can defend a 10-square kilometer area. The ECM system was developed by the

Northeast China Research Institute of Electronic Technology. See Liu Hsiao-chun, "Combat Effectiveness of China's Electronic Technology in Perspective—Causing the F-117 Stealth Fighters To Malfunction," *Kuang chiao ching*, July 16, 1998, pp. 96-98, in *FBIS-CHI*-2875-98. The organic Second Artillery ECM system is discussed in a very comprehensive account of developments within the PLA by Lin Chong-Bin (Chong-Pin Lin), entitled *Heba* (Nuclear Hegemony), p. x.

171. For one of the best Taiwan assessments of the challenges presented by the PLA's growing arsenal of theater missiles, see Teng Hsin-yun, "Another TMD Episode—The PRC Missile Threat and Our Countermeasures As Well As Blind Spots in Taiwan's Defense Thinking," *Chien-tuan k'o chi,* May 1, 1999, pp. 100-107, in *FBIS-CHI*-0872-99, June 6, 1999.

172. This "area denial" concept is discussed in Thomas G. Mahnken, "Deny U.S. Access?," *Proceedings*, September 1998, pp. 36-39. For another evaluation of the implications of increasingly accurate and lethal theater missiles, see Paul Bracken, "America's Maginot Line," *Atlantic Monthly*, December 1998.

173. The six specious arguments are: (1) TMD will cause an arms race; (2) TMD will contradict the ABM Treaty; (3) TMD will encourage Taiwan independence sentiment; (4) TMD can be used offensively; (5) TMD will lead to the militarization of space; and (6) TMD "violates" the Three Communiqués. All are based on oversimplifications and half-truths as seen in the following paragraphs:

(1) Arms races generally are caused by one side's rapid buildup in offensive capabilities. One could argue that an accelerated arms race has been underway in the Taiwan Strait since the early 1990s. Undercutting Beijing's overwhelming offensive advantage through viable defenses could enhance cross-Strait stability by raising the costs of using force.

(2) Questions surrounding the ABM Treaty only apply to *upper tier* systems—U.S. upper tier systems now being tested have been certified as ABM Treaty compliant.

(3) The argument that TMD will encourage Taiwan independence sentiment is also misleading. There are more important factors besides defenses that fan the flames of Taiwan independence. PRC policies that alienate Taiwan are most relevant. Besides, active missile defenses would not encourage independence sentiment any more than other weapon systems, such as fighters, surface-to-air missiles, or ships. One could also argue that Taiwan's indigenous capacity for defense is only a

minor factor influencing public sentiment regarding greater autonomy since, according to some sources, Taiwan's domestic polity is largely uninterested in defense issues.

(4) The argument that active missile defenses can be used offensively is also based on a half-truth. Converting upper tier interceptors to surface-to-surface missiles could enable strikes against targets at long ranges. However, such a means is not cost efficient due to payload limitations. Development of a cost effective theater ballistic missile is well within Taiwan's technical capabilities. The argument that missile defense systems incorporate technologies useful to ballistic missiles (i.e. gyroscopes and accelerometers) assumes that Taiwan: a) does not have the indigenous capacity to develop viable inertial measurement units; b) would be willing to violate MTCR assurances granted to the United States; and c) would take the trouble to reverse engineer the guidance technology.

(5) The argument that TMD could lead to a militarization of space is partially true. If supported by a robust search, acquisition, and tracking network, *upper tier* systems could be used to strike some satellites in *low earth* orbit.

(6) Finally, active missile defenses would not "violate" the Three Communiqués. First, the Three Communiqués are parallel statements of policy that have little standing in international law. Secondly, the argument that U.S. provision of active missile defenses would revive the U.S.-Taiwan defense alliance, undermining the foundation of U.S.-PRC relations as spelled out in the 1979 Communiqué, is based on the assumption that TMD would require operational connectivity (i.e. satellite early warning) with the United States. This is not necessarily true. While satellite early warning could enhance the effectiveness of missile defense systems, such as PATRIOT and THAAD, and can operate autonomously against SRBMs, early warning radar can supplant satellite early warning. The idea that satellite early warning necessarily equates to an alliance is also fallacious. The United States has several early warning arrangements with foreign governments that are not considered allies. Furthermore, missile defenses do not "violate" the 1982 Communiqué any more than other weapon systems. As Assistant Secretary of State John Holdridge pointed out in his August 1982 Congressional testimony, the U.S. agreement to reduce arms sales to Taiwan was contingent upon Beijing's peaceful approach to resolving the Taiwan issue, generally characterized by its military posture directed against Taiwan. As Holdridge pointed out, a rise in the military threat to Taiwan theoretically would be accompanied by a rise in U.S. security assistance, in accordance with U.S. domestic law under the Taiwan Relations Act.

174. Among numerous references, see Tom Plate, "East Asia, Infected By a New Arms Race, Risks Deadly Miscalculations," *LA Times*, July 7, 1999; and Vanessa Guest, "Missile Defense Is Wrong Call on Taiwan," *LA Times*, May 3, 1999.

175. Among numerous references, see Robert Jervis, "Offense, Defense, and the Security Dilemma," *World Politics*, Vol. 30, No. 2, January 1978, pp. 186-214; and Stephen Van Evera, "Offense, Defense, and the Causes of War, *International Security*, Vol. 22, No. 4, Spring 1998, pp. 5-43.

176. See Stephen Van Evera's Spring 1998 *International Security* article for a comprehensive argument on the dangers of an offense-dominated security environment. In line with this reasoning, one could argue that Taiwan's newly procured F-16s are to blame for intensification of the cross-Strait arms race. However, Taiwan's F-16 fleet is optimized for air-to-air operations, not long-range strike.

177. Hou Xiaoyan, "Taiwan zhongkeyuan de yanjiu linghuo he chanpin jieshao" (Introduction to Taiwan's Zhongshan Institute of Science and Technology Fields of Research and Products), *Feihang daodan (Cruise Missiles)*, December 1998, p 39. This CASC journal asserts that CSIST is converting the TK-II into a surface-to-surface ballistic missile. The ballistic missile, called Sky Halberd (*Tianji*), allegedly has a design range of 320 kilometers and a 150-kilogram warhead. CSIST is integrating GPS with their inertial navigation systems in order to achieve a CEP of less than 100 meters. Primary targets would be airfields, military ports, and industrial areas. The CASC author also asserts that a second stage addition to the *Tianji* missile would significantly expand the range. As an example of the nuclear weapons development debate, see Liao Hung-hsiang: "Should Taiwan Develop Strategic Nuclear Weapons?" *Ch'uan-ch'iu fang-wei tsa-chi*, March 15, 1999, pp. 18-21, in *FBIS-CHI*-0018-99, April 29, 1999; and Liu Chien-hua, "What Equipment Should Taiwan Use to Defend Itself, *"Tzu-li wan-pao,* November 9, 1997, p. 2, in *FBIS-CHI*-97-364, December 30, 1999. For background on Taiwan's previous nuclear weapons development program, see David Albright and Corey Gay, "Taiwan: A Nuclear Nightmare Averted," *The Bulletin of Atomic Scientists,*, Vol. 54, No. 1, January/February 1998. Also see Gerald Segal, "Taiwan's Nuclear Card," *Asian Wall Street Journal*, August 5, 1998.

178. Robert Jervis, "Offense, Defense, and the Security Dilemma," *World Politics*, Vol. 30, No. 2, January 1978, pp. 186-214.

179. For an outstanding discussion on shortcomings of air and missile campaign theory, see Colonel Richard Szafranski, "Parallel War and Hyperwar: Is Every Want a Weakness?" in Barry Schneider and Lawrence Grinter, ed., *Battlefield of the Future: 21st Century Issues*, Air War College Studies in National Security No.3, Air University, 1995, pp. 125-148.

180. See Teng Hsin-yun for a realistic Taiwan assessment of countermeasures.

181. Theoretically, assuming two interceptors are employed for every ballistic missile and a 100 percent probability of kill, a MADS battalion could engage a near simultaneous salvo of 48 SRBMs directed against targets within its area of coverage. For background on countering theater missile threats, see Joint Pub 3-01.5, *Doctrine for Joint Theater Missile Defense*, February 22, 1996. DoD's *Report to Congress on Theater Missile Defense Architecture Options for the Asia-Pacific Region* notes there are multiple options for active missile defenses. Twelve land-based lower-tier fire units could provide for partial coverage of Taiwan's most crucial assets. Eleven sea-based lower systems could cover the entire island. Neither lower tier system, however, can counter longer-range threats such as the DF-21. In addition, a maneuvering reentry vehicle that complicates lower tier engagements would drive Taiwan toward upper tier solutions. A single THAAD-like unit could engage all known missile threats. The exo-atmosphere Navy Theater Wide-like system can cover the entire island but would not be able to engage shorter range threats such as the 300-kilometer DF-11 since it would not leave the atmosphere.

CHAPTER 6

PLA AIR FORCE OPERATIONS AND MODERNIZATION

Kenneth W. Allen

Introduction.

> The Central Military Commission has called for the urgent upgrading of the country's Air Force to neutralize growing threats from regional neighbors and other countries. . . . Our country now faces a serious challenge. . . . China needs to develop airborne early warning systems and foster research in the development of high-tech electronic combat systems. . . . If threatened from the air, China must have the ability to carry its defense strike capability to targets outside its own airspace.
>
> *Jiefangjun Bao*, April 7, 1996[1]

> The Chinese Air Force plans to acquire state-of-the-art weapons systems by early next century, including early warning planes, electronic warfare warplanes, and surface-to-air missiles. The PLA Air Force is now able to fight both defensive and offensive battles under high-tech conditions.
>
> Liu Shunyao, PLA Air Force Commander, April 1997[2]

China's Air Force has significantly improved its combat readiness. During 1998, pilots achieved a record of per capita flying time, the highest since 1985, in spite of heavy summer flooding and a program to restructure the Air Force. Pilots paid particular attention to improving basic flying techniques. The fact that 66 percent of air units conducted highly successful long-distance mobile maneuvers under harsh weather conditions indicated that China's Air Force has greatly enhanced its combat readiness.

> Wu Guangyu,
> PLA Air Force Deputy Commander,
> January 1998[3]

We should build an Air Force capable of both offensive and defensive operations with Chinese characteristics.

> Jiang Zemin, March 1999[4]

China's Air Force, known as the People's Liberation Army Air Force (PLAAF), is in a crucial transition period, as it changes from an obsolescent giant to a modern force prepared to fight local, limited wars under high-tech conditions. The PLAAF is slowly moving from a defensive force dominated by 1950s vintage combat aircraft with short legs and limited all-weather intercept capabilities to an offensive-oriented force with extended range and greater lethality. While new aircraft like the J-10, J-11 (Su-27), and Su-30 are gradually introduced into the force, older aircraft like the J-7 and J-8 are being modified with better avionics and air-to-air missiles to bridge the gap. The new combat aircraft force of the 21st century will be controlled by airborne early warning aircraft, refueled by tankers, and supported by electronic countermeasure and intelligence collection aircraft. The PLAAF is forging ahead with advanced tactics and logistics techniques for its newer aircraft, while sustaining the operational capabilities of its older inventory. In addition to its combat aircraft, the PLAAF is improving its surface-to-air missile force and mobility for its elite airborne corps. In 10 years the PLAAF will be a much smaller force, but will have greater range and lethality than the PLAAF of the 1990s.[5]

PLAAF writings indicate that it has an impressive history defending China. The PLAAF established its credibility during the Korean War, the 1958 Taiwan Strait Crisis, and the Vietnam War, but it has not conducted any large-scale air battles since then. The official PLAAF history states that it has shot down 1,474 and damaged

2,344 aircraft of all types. Analysis of these figures shows that PLAAF aircraft have shot down or damaged only about 200 aircraft during air-to-air combat, most of which occurred during the Korean War. The PLAAF's antiaircraft artillery (AAA) and surface-to-air missiles (SAM) shot down or damaged the remaining 3,600 aircraft. The PLAAF's SAM forces are particularly proud of shooting down five Nationalist Air Force U-2 reconnaissance planes between 1963 and 1967.[6] The last PLAAF combat took place in October 1987, when a SAM shot down a Vietnamese MiG-21 that had crossed the border.

The PLAAF readily admits that its equipment is backwards, but modernization is taking place across the board, including equipment, tactics, training, logistics, and maintenance. Unlike the 1980s, there are very few PLAAF leaders left who fought in the Korean War. Today's Air Force leaders have put forth an optimistic vision of the future. Unfortunately, the PLAAF does not provide many solid clues as to how it plans to reach that vision. One of the biggest problems foreigners have researching the PLAAF is the lack of open source information. Even when information is available, it focuses on the vision, not on the process. Most Western writings on the PLAAF tend to focus on the acquisition of hardware, such as the Russian Su-27s, Su-30s, Il-76s, and S-300s.[7] These articles pay little attention to the "software" issues, including leadership, missions and organization, strategy and doctrine, personnel, support equipment, operations, training, logistics, maintenance, and C4I (command, control, communications, computers, and intelligence).[8]

This chapter will address the PLAAF's operational capabilities and modernization. The first part will provide statements by the PLAAF's commander, Lieutenant General Liu Shunyao, describing the PLAAF's past, present, and future. The next part will look at the PLAAF's missions and organizational structure, including the aviation, air defense, and airborne forces. The third part will assess the air activity that has occurred over the

Taiwan Strait in reaction to President Lee Teng-hui's July 9, 1999, statements about "state-to-state" relations between Taiwan and the mainland. The chapter will also address China's neighbors' perceptions of the PLAAF. The final part will provide conclusions.

PLAAF Commander's Assessment.

Lieutenant General Liu Shunyao has had several interviews with Chinese reporters since he became PLAAF commander in December 1996.[9] The first of these interviews coincided with Taiwan's receipt of the first group of 150 F-16 fighters from the United States and 60 *Mirage* 2000-5s from France. The interviews also came after the PLAAF had already received several Russian Su-27 fighters, Il-76 transports, and S-300 SAMs. The PLAAF had also deployed its first indigenous B-6 airborne refueling aircraft for its J-8II fighters, and had ordered the first Il-76 airborne early warning platform from Russia and Israel.

During the interviews, Commander Liu stressed that the aviation troops formed the Air Force's backbone. The PLAAF culture has always focused on the role of aircraft as the core of the force, even though the air defense (SAM and AAA) forces have shot down more aircraft. He also emphasized that the PLAAF's 15th Airborne Army, which has also become more integrated into the PLA's joint operations, has established "fist" units that are now more mobile and have longer range as a result of acquiring several Russian Il-76 transports.[10]

While discussing the PLAAF's current capabilities, Commander Liu has focused on the ability to fight defensive and offensive battles under high-tech conditions. The shift from strictly defensive to offensive capabilities derived from the PLAAF's post-Gulf War realization that precision guided missiles (PGMs) and long-range cruise missiles had changed the rules of warfare. This realization was one of the driving forces for the PLAAF to acquire modern weapon systems from Russia. As a result of modernizing its weapon systems, the Air Force has been able to changes its tactics, extend its combat range, and practice providing support for ground and naval operations.

Liu and other PLAAF officials continue to emphasize that three-quarters of its pilots are now able to fly in all-weather conditions; the percentage of category "A" combat regiments, an indicator of the Air Force's combat effectiveness, has reached 95.5 percent; and that the Air Force has a 16-year flight safety record.[11] As for flying in all-weather conditions, there is some question as to exactly what types of flying the pilots conduct under visual flight regulations (VFR) and night flying conditions. Are these types of flights flown only on cloudless, moonlit nights, or during cloudy, pitch-black conditions over long distances while engaged in air intercept training? Since the Chinese media rarely reports aircraft accidents, there is no way to verify the accuracy of the flight safety record, but this claim, also, is questionable. For example, there are credible reports that Cao Shuangming, the PLAAF commander from 1992-94, was relieved of duty partly due to a series of aircraft accidents that took place under his command. Furthermore, since the PLAAF rarely trains using rapid aircraft turn around sorties and most engines can only be used from 100 to 300 hours before they are overhauled, the maintenance record would probably be reduced considerably during periods of sustained use, such as during a conflict.

Given the PLAAF's current limitations, Commander Liu and other Air Force leaders have expressed an optimistic vision of the future. As the Air Force upgrades the capabilities of its current inventory with foreign technology, begins producing the J-10 and J-11 (Su-27) aircraft, and acquires new fighters (Su-30) and airborne early warning aircraft and SAMs, the PLAAF will be able to extend its range, increase its firepower, and change its tactics through the use of improved electronic warfare, night training, and joint campaign training. The PLAAF will also focus its efforts on research and development, while importing some high-tech weapons.

The debate in China about importing weapon systems versus domestically producing them continues. For example, a July 1999 report in *Science and Technology Daily* complained that China's achievements in many areas over the past 20 years have attracted world attention, but the aviation industry has increasingly lagged behind. The aviation industry continues to have a sprawling organization with weak facilities and low standards, and the gap between it and the aviation industries of the developed western countries is widening. The article lamented that this is cause for anxiety, because without a strong defense there can be no genuine overall strength.[12]

There is also a debate among foreign analysts about how much China's aviation industry has been able to learn from foreign aircraft manufacturers and transfer the knowledge to producing combat aircraft. Foreign aircraft manufacturers have routinely complained for almost 2 decades that they have spent time and money training Chinese technicians to work on a specific commercial aircraft project, only to have those technicians disappear to some unknown project and be replaced by a new batch of trainees.

While some observers see China's aviation industry as large and stagnant, other observers cite some centers of excellence. Facilities such as the Chengdu Aircraft Company, Xian Aircraft Company, and Shenyang Aircraft Company have made significant progress through joint ventures with Western companies and assistance from Russia and Israel. For example, Boeing 737 tail sections are produced in China with no U.S. backup. These ventures have provided both access to Western manufacturing processes and valuable technical and managerial training. While these are derived from commercial aircraft manufacturing, the processes are generally the same for a mostly metal fighter aircraft like the J-10 and the Su-27.

Only time will tell how these aircraft production programs will turn out. Because of the size of China's air forces, however, any modernization plan will require Chinese production to be economically feasible.

As the PLAAF upgrades its weapons systems, it must also modernize its organizational structure, tactics, training, and support capabilities. One problem associated with analyzing the PLAAF is determining which changes are being tested and which changes are being implemented. PLA journals often discuss research or changes that have taken place at the unit level. Some analysts, however, assume that these changes have been, or will be, implemented throughout the force. This is not necessarily the case. The PLAAF, like the PLA as a whole, has a complex method of introducing new ideas, weapon systems, and tactics into the force. The General Armaments Department has attempted to streamline this process, but only time will tell whether or not it is successful.

After identifying a problem, the PLAAF systematically seeks a solution. First, the appropriate research institute and/or the PLAAF Command College studies the problem from a theoretical perspective and makes recommendations.[13] Next, PLAAF Headquarters approves the recommendations. After this, the theories are tested at the unit level. Occasionally, separate units test different alternatives simultaneously. Based upon evaluations, competing theories may be accepted, modified, or rejected. After the competing theories are tested and accepted at the unit level, they are tested at the next level, and so forth up the organizational structure. At some point, the PLAAF selects one of the theories and begins implementing it force-wide, starting again at the unit level.

This is a time-consuming process and involves a tremendous amount of coordination. The problem becomes even more complex if the theory being tested involves more than one PLAAF branch or PLA service. Whereas a change in flying tactics involves only the PLAAF's aviation branch, changes in the logistics system to support joint service operations involves the entire PLA. For example, the 1995 and 1996 exercises opposite Taiwan were intended as a show of force, but they also provided the PLA with the opportunity to test joint training that had been evolving since the end of the Gulf War.

Doctrine and Strategy.

The PLAAF's evolution of operational capabilities is tied to the evolution of the PLA's overall doctrine and strategy. China's communist leaders have long seen themselves as encircled by real or potentially hostile forces threatening the regime's security. They have also long sought to define a doctrine and strategy to deal with this situation. The PLA's doctrine has evolved from Mao Zedong's basic doctrine of people's war, which still retains a measure of influence in Chinese thinking, at least in broad conceptual terms. In 1985, the CMC radically revised China's doctrine and strategic defense policy by directing the armed forces to change from preparation for an "early, major, and nuclear war" to preparing for "local limited wars around China's borders, including its maritime territories and claims." Following the Gulf War, this doctrine was amended to "fight local wars under modern, high-technology conditions." Thus, people's war has evolved as a blend of defense and offense, and has been modified to incorporate various strategies, including active defense and the rapid-reaction strategy.[14]

In its essence, the people's war doctrine reflects a strategy of weakness. Since the PLA was founded in 1927, it has had to develop strategies for defeating adversaries who had superior weapons and equipment. That this dilemma should continue into the 21st century is no doubt frustrating to members of the current military hierarchy.

While people's war places special emphasis on defensive strategy and on the factor of manpower over weapons, the PLA has never ignored the need for an offensive strategy.

The PLA's involvement in the Korean and Vietnam wars, plus its attacks against India in 1962 and Vietnam in 1979—which were followed by quick, unilateral withdrawals—were all described as defensive operations. At the same time, Mao long recognized the value of utilizing superior force to overwhelm China's adversaries. Within the people's war doctrine, the basic military strategy Mao formulated, known as active defense, was one of a protracted, defensive war.

In a recent paper, Dr. Paul Godwin states that,

> the PLA has been shifting over the past 20 years from continental defense in depth to peripheral defense and maritime force projection, and from a ground-force dominated approach to war, to a multi-service joint operations doctrine. In conceptualizing the battlefield, the PLA has shifted from a two-dimensional concept, where the ground war was the central focus, to a multidimensional battlespace, where space and cyberspace play roles as important as the traditional air-land-sea dimensions. The PLA has faced the major difficulty of the absence of any period of stability in which it could complete the organizational, training, and logistics changes required to implement a revised strategy and operational doctrine.[15]

Although PLAAF writings mention the broader PLA doctrine and strategic concepts of people's war, people's war under modern conditions, and people's war under modern, high-tech conditions, they tend to focus more on campaign strategy, campaign tactics, and tactical training.

As an arm of the PLA, the PLAAF traditionally has conducted its combat operations as a series of subordinate campaigns within the PLA's overall campaign. The PLAAF describes a campaign as "using from one to many aviation, air defense, or airborne units to carry out a series of combined battles according to a general battle plan to achieve a specified strategic or campaign objective in a specified time."[16] During March 1997, Commander Liu stated the PLAAF must improve its capabilities in actual combat by highlighting campaign and tactical training. He emphasized that campaign training involves air deterrence, air interdiction, air strikes, and participation in joint exercises.[17]

Although the PLA has always had an active defense strategy, one of the PLAAF's most significant developments in the past couple of years has been the public emphasis by Chinese leaders, including CMC Chairman Jiang Zemin, on the PLAAF's capability to fight offensive battles. What this means is that the PLAAF is beginning to acquire the types of weapon systems, such as the Su-27 and Il-76, that will allow the PLAAF to change its doctrine appropriately and move away from its purely defensive missions.

PLAAF Rapid-Reaction Force.

The PLAAF's Airborne Troops. One of the PLAAF's most important changes in campaign strategy took place in 1992, when the Air Force's 15th Airborne Army began changing into a rapid-reaction force (RRF). Although PLAAF airpower discussions in the late 1980s included ideas about fist units, these discussions centered on the airborne forces and not the aviation units. While the airborne forces were clearly included in plans for the RRF, it appears that the airborne forces did not actually form any operational RRFs until around 1992.[18]

According to an October 1993 *Jane's Defence Weekly* report, China was in the process of changing the 15th Airborne Army's three brigades into divisions, in order to boost their rapid-response power. The 43rd Brigade, based in Kaifeng, was the first brigade to undergo expansion to a division. The other two brigades, the 44th at Yingshan and 45th at Huangpi, followed suit shortly thereafter. Military planners had apparently decided that brigade-size forces were too small for their assigned combat missions. Chinese brigades normally have about 3,000 to 4,000 troops, and ground force divisions have about 15,000 troops.[19] The airborne units are composed of eight types of troops: scouts, infantry, artillerymen, signalmen, engineers, antichemical warfare corps, and automobile corps.

The 1999 Department of Defense's (DoD) assessment of the PLA describes the 15th Airborne Army as consisting of three airborne divisions, each with about 10,000 troops. The 15th Airborne Army is China's primary quick reaction force and has been designated as a strategic rapid reaction unit, but the airborne units remain handicapped by insufficient airlift. Acquisition of additional aircraft and modern equipment, together with the increased emphasis on utilizing airborne forces during training exercises, would marginally improve the airborne army's combat capabilities.[20]

According to Commander Liu, however, since the PLAAF began receiving several Russian Il-76 transports in 1992, the airborne troops now have all-terrain, all-weather, omni-directional combat capabilities.[21] In order to adapt to various adverse operational conditions, the airborne units have conducted exercises in the snowfields of the Greater Khingan Mountains (Da Xinganling), the hot jungles on the Shiwan Mountains in Guangxi, and the Kunlun Plateau, located 4,600 meters above sea level where the air is thin.[22]

Airborne troop training over the past few years appears to have focused primarily in and around Tibet. At the same time, however, some airborne training has also concentrated on a Taiwan scenario. By reporting this type of activity, the government is apparently trying to send a signal to inhabitants of Tibet, Taiwan, and Xinjiang that the airborne forces are preparing for internal contingencies, should the need arise.[23] Reporting of these types of exercises also points out some of the airborne forces' limitations. For example, during the 1996 military exercise opposite Taiwan, the PLAAF inserted a small contingent of airborne troops onto Haitan island, but this portion of the exercise was scaled down due to inclement weather, again calling the reliability of Liu's statement into question.[24]

Aviation Rapid Reaction Force Units. Serious discussions about establishing aviation RRF units did not occur until around the early 1990s, which coincided with China's purchase of the first Russian Su-27s and Il-76s. According to interviews with foreign air force officers in Asia, the PLAAF's rapid-reaction aviation units are currently composed of J-7, J-8, and Su-27 fighters and Il-76 transports. Although the bulk of the PLAAF consists of J-6 fighters, the RRF does not include any of these aircraft.[25]

The PLA's rapid-reaction strategy is based on the premise that China will only be engaged in local wars for the foreseeable future, and that the PLA must strike quickly to end the war and meet Beijing's political objectives. Furthermore, cost is a big factor, since equipment is becoming more expensive and the cost of maintaining older weapon systems is rising.[26]

PLAAF Missions and Organizational Structure.

Over the past 50 years, the PLAAF has endeavored to undertake an exceptionally broad range of operational missions. The first mission the CMC assigned to the PLAAF in 1949 was the air defense of Beijing and Shanghai against Nationalist air raids. This mission expanded to include northeast China during the Korean War and to the southeast provinces during the 1958 Taiwan Strait crisis. Although Western writings normally refer to air defense as including aircraft, AAA, and SAMs, the PLAAF makes a clear distinction between its aviation troops (aircraft) and air defense (AAA/SAM) troops.

Today, the PLAAF still describes its primary mission as defending China's territorial airspace. However, this mission can best be described as defending China's major cities and industrial areas, which can clearly be seen by looking at the location of the PLAAF's airfields, combat aircraft, SAMs, and AAA. As the PLAAF acquires aircraft and SAMs with longer ranges, these envelopes will gradually expand. Although the PLAAF states that its secondary mission is supporting ground and naval forces, it has never successfully carried out direct support to the ground troops and officially states that it can only support them indirectly. The PLAAF still describes a joint exercise as having aircraft airborne at the same time that its own SAM and AAA units or that some ground force units are active in a different area. It is also questionable just how much the PLAAF can actually support the ground and naval forces in the future. The PLAAF lacks the proper aircraft and joint experience to accomplish close air support or interdiction and has only recently made nascent inroads into the overwater training necessary for naval air support. Support for the naval forces is left up primarily to the naval air force.

Published PLAAF sources also refer to informal, tertiary missions, such as assisting socialist construction, providing air services for disaster relief and air rescues, and artificial rainmaking support for farmers.[27] The PLAAF also supports other maritime activities. For example, according to a May 1998 *Xinhua* report from Guangzhou, the PLAAF was to conduct inspection flights in the coastal waters off Guangdong Province, as well as the Zhujiang River delta area, in order to enforce the May to September fishing ban. Guangdong Province and the South China Sea Branch of the National Bureau of Oceanography had requested the inspections in order to counter serious problems in excessive, unpaid, and disorderly utilization of maritime resources.[28]

In the absence of broader PLAAF mission statements, one must look at the PLAAF's organizational structure, operational branch functions, force locations, weapon systems, and planned weapons acquisitions, in order to analyze the PLAAF's full range of missions.

Administratively, the PLAAF's chain-of-command is organized into four levels: headquarters air force; seven military region air force (MRAF) headquarters; air corps and command posts; and operational units.[29] Headquarters

Air Force is equivalent to the U.S. Air Force's Air Staff and is organized administratively into four first level or major departments—headquarters, political, logistics, and armament—and their subordinate elements (second level departments, bureaus, divisions, offices, and sections).[30] The PLA's military region (MR) headquarters is responsible for combined operations, and the MRAF commander, who is also an MR deputy commander, is responsible for flight operations within the MR. The seven military regions are Shenyang, Beijing, Lanzhou, Nanjing, Guangzhou, Jinan, and Chengdu. Each echelon below Headquarters Air Force from the MRAF headquarters to the lowest level in the chain-of-command basically mirrors this administrative structure.

In order to perform its operational missions, the PLAAF is organized into branches—aviation, AAA, SAM, radar, and communications. The PLA's airborne troops belong to the Air Force, and sometimes, but not always, are noted as the PLAAF's sixth branch. The PLAAF also has schools, logistics units, repair facilities, research institutes, hospitals, and sanitoriums as part of its organizational structure. The Air Force's Logistics Department has its own water transport craft and boat troops to ship fuel to PLAAF units along the Yangtze River and coast. The PLAAF's political structure ensures Party control at all levels.[31] By looking at modernization of the PLAAF's six branches, it is possible to understand more fully the scope of the Air Force's operations and capabilities.

The PLAAF's Aviation Branch. The PLAAF was formed around its aviation troops (*hangkongbing*), which remain the Air Force's main arm. The aviation troops are organized into fighters, ground attack aircraft, bombers, transports, and reconnaissance aircraft. These aircraft are organized into air divisions, regiments, groups, squadrons, and associated maintenance, logistics, and support units. The support units are organized into regiments, battalions, and companies. There are also independent air regiments and groups, which conduct specialized missions, such as operational test and evaluation of equipment, reconnaissance and surveying, troop transport, and reforestation. For the most part, these special mission aircraft include reconnaissance fighters and Il-14 and Yun-5 transports. Air divisions can be directly subordinate to Headquarters Air Force, an MRAF headquarters, an air corps, or a command post.

A typical air division headquarters consists of the command staff and administrative organization. These organizations are responsible for combat training, political training, supply, and maintenance support for the division. Each division and regiment has a Party committee and a standing committee, of which the political commissar serves as the secretary for both committees. The division's Party committee consists of the standing committee, which includes the division's command staff, plus the commanders and political commissars of each subordinate regiment.

Because the PLAAF has historically been synonymous with the aviation troops, the Headquarters Department's second level Operations Department has basically functioned as the "Aviation Troop Department." The Operations Department also is responsible for the airborne troops. This is in contrast to the separate second level departments that have been established for the radar, communications, and AAA/SAM branches.[32] This separation of aviation and air defense permeates the entire chain-of-command, keeping administrative firewalls between the aviators and the rest of the PLAAF.

Origins of the Aviation Branch. When the PLAAF was established, its aviation troops were organized into several brigades with three to four regiments each. For example, the 4th Combined Brigade was established at Nanjing in June 1950 and became the PLAAF's first aviation troop unit. It consisted of 10th and 11th Pursuit Regiments, the 12th Bomber Regiment, and the 13th Attack Regiment. By the end of 1950, these brigades reduced the number of regiments to two, dropped the type of unit (Pursuit/Attack/ Bomber) from the name, and became air divisions.[33]

Since 1951, the number for regiments per division has fluctuated between two and three, depending upon the number of aircraft available and the changing missions. Some flying academies have up to four regiments. By the end of May 1951, the PLAAF had 17 air divisions, each with two regiments, including 12 pursuit divisions, 2 attack divisions, 2 bomber divisions, and 1 transport division. This expanded rapidly, so that by March 1953, a total of 28 air divisions and 56 air regiments were formed. At the same time, each division began changing from two regiments back to three. By early 1954, the PLAAF had 28 air divisions and 70 air regiments, with 3,000 aircraft.

From 1960-65, more air divisions were created to guard the coast. From 1966-76, aviation troop units were expanded to cover the rest of China. Although the PLAAF had 50 air divisions by the mid-1980s, the number was reduced to around 45 by 1990 as older aircraft were gradually taken out of the inventory.[34]

According to the 1999 DoD report, the PLAAF currently numbers over 400,000 personnel with approximately 4,500 combat aircraft organized in some 30 air divisions. The PLAAF also maintains about 150 transport aircraft in two air divisions.[35] A PLAAF air division can have one or more air regiments per airfield, with each airfield assigned a field station for logistics support. Although a division can have more than one type of aircraft (i.e., J-7s and J-8s), each regiment typically has the same type of aircraft. The table of organization and equipment (TO&E) for a typical air regiment consists of from 25-32 aircraft, but may actually have more or less assigned. The regiment is the basic organization for training and operations. Each regiment has three flying groups and one aircraft maintenance group. Each flying group is further divided into three squadrons.

The division has about one and one-half to two pilots per aircraft. Although the pilots are assigned to squadrons, each with three to five pilots, the aircraft are assigned to the regiment as a whole, not just to the squadrons. Each pilot, however, normally only flies one to three airframes, so they become familiar with each aircraft's handling capabilities. The average pilot remains in the Air Force until he or she retires.[36] The PLAAF established age limits for its pilots in the 1980s: fighter and ground-attack pilots, 43–45 years; bomber pilots, 48–50 years; transport pilots, 55 years; helicopter pilots, 47–50 years; and female pilots, 48 years. The average fighter and ground-attack pilot is 28 years old.

Flight Training.

To help build *esprit de corps*, the PLAAF began awarding four pilot ratings in 1986—special, first, second, and third grade—which are awarded after the pilots complete their initial and upgrade training at an operational unit. The criteria includes time-on-station, flying hours, "flying in weather," and special missions. In addition, the PLAAF awards aircrew ratings to navigators, communications and gunnery personnel, and instructor pilots. According to a 1989 PLAAF report, of the 10,000 pilots in the entire PLAAF at that time, 7 percent of the total and 15 to 20 percent of the fighter pilots were special grade.[37] There are differing reports about the educational background of today's PLAAF pilots. In April 1999, Commander Liu stated that all air division and regiment leaders are special-grade or first grade pilots, and one-half of the pilots in the flight units are college-educated.[38] This is in contrast to an interview in 1989 with then-PLAAF Political Commissar Zhu Guang, who stated that all the pilots had a college education or above, and to a 1999 article by John Lewis and Xue Litai that states,

> In 1997, the PLAAF finished drafting its training programs for high-tech wars, but in carrying them out, it has encountered a fundamental problem because only 20.7 percent of the air officers are college graduates. Quick fixes or short-term training classes cannot solve the lack of qualified technical

personnel to operate high-tech air weapons in an environment that attracts the best to civilian occupations.[39]

There are no ready explanations for this apparent drop in college educated officers, especially since all PLAAF officers must attend a PLAAF 4-year academy, many of which now offer post-graduate degrees as well.

Based upon an analysis of Chinese literature and interviews in China, it is evident that PLAAF pilots do not fly as many hours as their Western counterparts. According to interviews with PLAAF and foreign air force officials, the PLAAF's flying hours have not changed appreciably over the past 15 years, but they have changed their training techniques. Since the end of the 1970s, bomber pilots have consistently flown an average of 80 hours per year; fighter pilots 100 to 110 hours; and A-5 ground attack pilots up to 150 hours.[40] This compares to about 215 hours per year for U.S. Air Force bomber, fighter, and attack crews. USAF pilots also conduct numerous hours training on advanced simulators.[41]

Based on interviews in Asia, it appears that PLAAF pilots are flying the Su-27s between 60-100 hours per year, and supplementing this with flight time in J-7s. These interviews also indicate that every country considers the PLAAF's Su-27s as their primary concern, regardless of how many hours the pilots train. The acquisition of these aircraft has definitely had a significant psychological impact on China's neighbors.

Since 1996, PLAAF pilots have been noted flying in more sophisticated simulated air-to-air combat with newly-formed aggressor units, employing jamming, flying over the Taiwan Strait, conducting live missile firings over the East China Sea, dropping parachute-retarded bombs at a bomb range. There are also Chinese writings that the PLAAF conducts post-mission analysis of aerial combat using information from flight data recorders that contain information from the air data computer and possibly some fire control data. The Chinese are also seeking Air Combat

Maneuvering Instrumentation (ACMI) pods similar to those used by Western forces. These pods send information to ground personnel so they can evaluate aerial engagements on a real-time basis.[42]

Although the PLAAF may not fly as many hours as the USAF, the PLAAF believes that its training is improving and is adequate to conduct its missions. Some Air Force leaders firmly believe that their intelligence, mobility, and attack capabilities will be sufficient to allow them to react appropriately to any situation, including gaining air superiority, supporting the ground forces, and conducting counterattacks against targets inside the enemy's borders. Furthermore, interviews with foreign air force officials indicate that they believe Western reporting does not take into account how the PLAAF matches up against China's neighbors. For example, Commander Liu noted that aviation units during 1996 exceeded their annual training plan requirements by 1.8 percent. Highlights during that time included night flying, live bombing and missile firing, training over the ocean, low-altitude flights, and emergency mobility deployments to other airfields. This type of training accounted for 45 percent of the planned annual training time.[43] What the article failed to state was that most of this training took place in a single exercise opposite Taiwan, and that several portions of the exercise were curtailed due to inclement weather.

To round out the PLAAF's tactical training and help make up for the limited number of flying hours per year, the PLAAF has increased its use of flight simulators. The PLAAF now reportedly conducts more than 90 percent of its tactical and technological training on simulators. The PLAAF has developed a full-spectrum spherical screen simulator, three-dimensional flight simulator, and in-flight refueling plane simulator for pilots to train and develop special flight and air strike skills under simulated combat conditions. Simulation capabilities have evolved from electromechanical simulation to laser, electronic, and computer simulation; from technical simulation to tactical and campaign simulation; and from the simulation of a single armament or aircraft type to integrated simulation of the main battle arms combined with multiple aircraft types and various forms of weaponry.[44] If these types of simulators have, in fact, become operational at all aviation units, then they will definitely help supplement pilot training, which is minimal at best. During the 1980s, the Air Force only had a few nonstandard simulators. Most of the pilots trained on an instrument board—a mock cockpit with a wooden panel containing all of the cockpit's instrument dials on it.

In January 1999, PLAAF Deputy Commander Wu Guangyu emphasized the PLAAF's increased readiness by stating,

> China's Air Force has significantly improved its combat readiness. During 1998, pilots achieved a record of per capita flying time, the highest since 1985, in spite of heavy summer flooding and a program to restructure the Air Force. Pilots paid particular attention to improving basic flying techniques. The fact that sixty-six percent of air units conducted highly successful long-distance mobile maneuvers under harsh weather conditions indicated that China's Air Force has greatly enhanced its combat readiness.[45]

Commander Liu also discussed training reforms by stating that several units have been selected to carry out a series of reforms on tactics. Those units have tried, demonstrated, and refined a series of advanced combat theories and propositions in live-fire training exercises. They have scored excellent initial results in mobile operations, air strike, air superiority, air control, night attacks, and defensive operations.[46]

PLAAF Aggressor Unit. The statements above most likely describe the PLAAF's "Blue Army" aggressor units.[47] According to several *Jiefangjun Bao* articles and interviews with PLA and non-Chinese air force officials, the PLAAF has established an aggressor unit at its Flight Test and Training Center at Cangxian to simulate offensive and defensive operations against the "Red Army."[48] An April 1997 *Jiefangjun Bao* article reported that,

> The Blue Army aggressor unit is equipped with advanced equipment and flown by special-grade and first-grade pilots. This unit has fought numerous air battles with every PLAAF combat aircraft unit, in order to help improve the PLAAF's high-tech combat effectiveness and improve their knowledge of tactics. The specialized unit has emerged as a strong enemy force in simulated air battles, which has led the PLAAF to make numerous changes in its tactical training. In light of China's training situation, the specialized unit has launched no-notice preemptive strikes under real combat conditions.[49]

The air tactics concepts that the Cangxian training center has developed against the aggressor unit are reportedly now being moved to the unit level, where several units have begun to turn these concepts into live-fire exercises. According to the April 1997 *Jiefangjun Bao* article, the PLAAF has obtained some initial results in important combat study areas, such as maneuverable combat, air attack, fighting for air supremacy, night attack, and defense. As a result of this training, the PLAAF has adopted a new training syllabus characterized by adaptability to combat situations under future high-tech conditions.

According to various interviews, similar aggressor units have been established in each of the seven military regions. The aggressor aircraft engage in exercises with local units, employing dissimilar aircraft air intercepts utilizing evasive maneuvers.[50] Although interviews indicated that the aggressor unit consists primarily of J-7 and J-8 interceptors and does not have any Su-27s assigned as of yet, this situation may have changed. A recent news article from Hong Kong described an exercise where,

> Four Blue Army Su-27s, simulating Taiwan Air Force F-16s, flew from an undisclosed airfield in North China [probably Cangxian] and conducted air-to-ground attacks against enemy targets. The report noted that the aggressor Su-27s

flew a total of 1,600 kilometers, with two SU-27s conducting strikes on a Red Army missile position, a command post, and two radar positions. The Su-27s also destroyed eighteen J-7s parked on the airfield. The remaining two Su-27s flew cover for the other two aircraft and shot down several J-7s in air-to-air combat.[51]

If the aggressor squadron now has Su-27s permanently assigned, this is a significant step forward in tactical training for the PLAAF.[52] The article also points out, however, the J-7s weaknesses in conducting their air defense mission. Whereas the Su-27s radar system and AAMs are capable of shooting down aircraft beyond-visual-range (BVR) in a head-on aerial engagement, the J-7s and older J-8s do not have either capability. The J-7s and J-8s must be within visual range and do not have a head-on intercept capability.

As a result of this more aggressive training program, PLAAF pilots have intensified their training under different weather conditions, at lower altitudes, and, most significantly, over water.[53] They have also practiced rapid deployment to permanent and auxiliary airfields. The PLAAF has also improved its navigation capabilities. There is a density of radio navigation aids (Tactical Air Navigation [TACAN] for the J-7 and J-8 and Short Range Radio Navigation System [RSBN] for the Su-27) available within China. In addition, the Chinese have indicated that they have retrofitted some of their aircraft with global positioning systems (GPS). Furthermore, the Su-27 and some J-8s have an inertial navigation system (INS) and can use their radar to navigate using islands and coastlines.[54]

In conjunction with these operational changes during the 1990s, the Air Force has had to adjust its logistics, including spare parts acquisition, storage, and distribution, and maintenance practices to meet these new challenges. These adjustments include computerizing individual logistics and maintenance operations, and then networking the computers within the unit and among different units at the same and higher levels. It has also meant establishing small logistics and maintenance teams capable of deploying by rail or air at a moment's notice to accompany the unit's aircraft deployment.

On several occasions, many PLAAF units have formed offensive and defensive exercise teams with Navy, Army, and Air Force AAA, radar, and SAM troops to conduct exercises modeled on future warfare. One *Xinhua* article stated that the PLAAF has generally considered flying low-altitude bombing raids over the ocean as dangerous, since the water and sky look almost the same to the pilot. According to the article, pilots have overcome this difficulty and have now flown large bomber groups less than 100 meters above the sea, hitting all the intended targets. These pilots are now experimenting with new tactics, such as flying close to strategic points at sea, launching surgical air strikes against enemy troops, waging offensive air campaigns, conducting over-the-horizon air combat, and imposing air and sea blockades.[55] These types of missions are new for the PLAAF, since the PLA Naval Air Force previously was responsible for support to the Navy and overwater intercepts of hostile targets.

PLAAF's Gobi Desert Training Center. In 1958, the PLAAF built a large center for testing its AAMs and SAMs in the Gobi Desert near Dingxin, Gansu Province.[56] During the mid-1990s, the Air Force began expanding this base to include a tactics training center, where multiple PLAAF units could practice the tactics developed at Cangxian and tested in individual units throughout the force. An October 1996 *Jiefangjun Bao* article described the training center as follows:

> The Air Force recently established a modern, comprehensive air tactical training base marked by combat training under actual combat conditions. The training base has air and ground tactical training ranges, simulated runways built to scale, a SAM base, AAA positions, radars, radar support vehicles, simulated enemy command posts, ammunition depots, and oil depots, which look real. A large number of simulated tanks are also deployed in combat positions.[57]

The article also described the training base's command and control center.

> Training directors and air division commanders receive air combat reports, direct air battles, and communicate with combat planes in the air. They monitor units in the exercise, and direct unit deployment and movement in a timely fashion through use of the base's advanced facilities. A monitoring, control, and appraisal system installed in the base's command center receives information in a timely manner on each plane's flight path, course, speed, altitude, and other parameters. The system monitors the activity through video recording, radar, and flight orientation systems, so training directors will have accurate information to evaluate the training results."[58]

The training center probably includes the mock up of a Taiwan Air Force airfield. According to an April 1999 report in Taiwan's *China Times Express*, "China has built a military airfield near Dingxin airfield in northwestern Gansu Province, that is identical to the Taiwan Air Force's Chingchuankang (CCK) airbase in central Taiwan, so the PLAAF can practice bombing the island."[59] Building mock enemy airfields is not an uncommon practice among air forces of the world, and provides realistic training opportunities.

Following the PLAAF's participation in the joint service exercise near Taiwan in early 1996, the PLAAF conducted a two-day, large-scale, offensive-defensive exercise utilizing its aggressor unit at the Gobi Desert training center during September 1996.

> The exercise directors made it clear from the beginning that this exercise was to be conducted under unknown conditions. Almost all of the exercise's contents were new to the PLAAF participants, including using dissimilar aircraft in coordinated, joint offensive-counteroffensive attacks against ground targets and in air-to-air combat under electronic warfare conditions. The Air Force conducted the exercise under the principle that none of the participants would receive prior information about the enemy's deployments, combat tasks, battlefield targets, or flight routes during different stages of the exercise. The

exercise, which was almost canceled because one of the participating aircraft crashed en route, included aerial combat to seize air superiority, bomber and ground attack aircraft strikes on targets after avoiding enemy intercepting planes, and airborne landing and anti-airborne landing operations.[60]

This report provides a valuable glimpse at the PLAAF's combat philosophy. Specifically, the directors assumed that they would know very little about the enemy just before or during the battle, due to the lack of real-time reconnaissance capabilities. Therefore, the attack must be planned on insufficient data. Yet another *Xinhua* report in 1998 stated, "During an early 1997 live-fire exercise involving various types of planes, command posts of eight air divisions explored tactics and conducted training on the base's advanced training systems and facilities."[61] Although the number of aircraft involved in these exercises was not specified, the PLAAF's philosophy has been to deploy overwhelming numbers of aircraft to a theater. For example, during the 1979 border conflict with Vietnam, the PLAAF deployed approximately 700 aircraft to the border area. However, while air force officials in Taiwan speculate that the PLAAF would attack Taiwan with waves totaling several hundred aircraft, there are no indications in PLAAF writings that it routinely trains with more than 4-8 aircraft in the air at a time.

Upgrading the Guangzhou MRAF. The PLAAF has designated certain military regions for the most rapid modernization and training. According to a July 22, 1999, *Guangzhou Yangcheng Wanbao* article, the Guangzhou MRAF has conducted advanced training with all of the PLAAF's branches. The PLA apparently chose to highlight these types of activities in the Guangzhou MR as a warning to Taiwan. This article appeared shortly after Taiwan's President Lee Teng-hui announced his "state-to-state" theory on July 9.

Reportedly, all of the Su-27 and J-8II fighter pilots have met the requirements for all-weather flights and for flying the lead plane of a four-aircraft formation; the B-6 bomber unit has had excellent results in long-range bombing raids; the Il-76 and Y-8 transport unit has notably enhanced its long-range air transport and combat capability; and the SAM, AAA, and radar units enhanced their air defense net.[62] The article further detailed the specific types of training the fighter and bomber units conducted.

> The fighter units successfully conducted tactical drills, including over-the-horizon air combat, multi-target attack, electronic countermeasure warfare, and offshore live ammunition firing. The bomber unit conducted training for the first time involving multi-wave, multi-target, all-weather, long-range, live ammunition bombing. Fully loaded with bombs, cells with multiple aircraft [probably eight aircraft] took off before dawn and flew over half of China to conduct timely and accurate bombing at an unidentified range in northwest China [probably the Gobi Desert training center]. From the desert range, they then flew to an island off the southeast coast to attack targets there. Upon completion of their bomb runs, they returned home in a rainstorm, landing after dusk.[63]

The Guangzhou report provided additional details about the PLAAF's airborne, SAM, and AAA training as follows:

> For the first time, the Guangzhou MRAF transport unit conducted a 6-hour training flight in complicated weather conditions late at night. The aircraft also conducted airborne exercises in an unfamiliar highland cold zone without ground command, ground marks, and weather information. The SAM unit became combat capable in less than one year after receiving new [unidentified] equipment in 1998. In June 1999, the fully armed SAM unit deployed by vehicle to the Gobi Desert training center in northwest China. The exercise involved multiple goals, including "dismantling, leaving, advancing, and launching" as well as "food, shelter, management, and storage." While at the training range, the unit employed its new missiles to attack low-altitude, close-range targets in an electronic warfare environment for the first time, and hit every target.[64]

Weapon Systems. The PLAAF's weapons systems are a combination of old, modified, and new aviation (aircraft) and air defense (SAM and AAA) systems. Since the early 1990s, the PLAAF has retired over 1,000 of its older combat aircraft, storing them at Rugao airfield in central China.[65] According to the U.S. Department of Defense report on the Taiwan Strait balance,

> The PLAAF inventory includes over 2,200 obsolete J-6/*Farmer* fighters (last produced in 1979), several hundred J-7/*Fishbed* and J-8/*Finback* fighters, and over 40 Su-27/*Flankers*. In addition, the PLAAF has 500 A-5/*Fantan* (modified MiG-19) ground attack aircraft and about 500 bombers, including B-6/*Badger* and obsolete B-5/*Beagle*. Both its aerial refueling and airborne early warning (AEW) programs are behind schedule, as are several of its indigenous aircraft development programs. By 2005, the PLAAF will possess nearly 2,200 tactical fighter aircraft, 500 ground attack aircraft, and 400 bombers, as older aircraft are retired. The majority of the mainland's air fleet still will be composed of second and third generation aircraft augmented by a limited number of fourth generation platforms.[66]

Fighters. By 2010, the PLAAF's fighter force will most likely consist of between 1,500 and 2,000 aircraft, with almost the entire J-6s and early model J-7s retired. The remaining force will consist of modified J-7s and J-8s.[67] These aircraft will initially be complemented by and then replaced by the J-11 (Su-27), J-10, and Su-30. The J-10 was developed to replace the J-6, J-7, and A-5. The Su-27 is an air superiority aircraft, replacing the J-8 in this mission. The Su-30 is being purchased for deep strike/interdiction and air superiority, much like the U.S. F-15E, and will replace the older B-5/*Beagle* bombers. All of these newer aircraft will have greater range and payload, and will not replace the older aircraft on a one-for-one basis.

There are differing opinions about the long-term viability of various indigenous programs, such as the J-7, J-8, and J-10 program, all of which have been considered for elimination at one time or another since the mid-1970s. During the 1970s and 1980s, China had no alternative to domestic production, these programs had a history of important bureaucratic support, and hundreds of millions of dollars had already been spent on the programs. Therefore, they continued through the 1980s at a moderate pace, while China began searching for foreign aircraft. Beginning in the mid-1980s, the aviation ministry opted to provide significant upgrades to these aircraft using domestic and foreign subsystems and foreign technical assistance.[68]

Over the past decade, some analysts have had doubts whether the J-10, which is under development in Chengdu, will eventually be operationally deployed, while others have predicted that the J-10 will become one of the PLAAF's most important fighters in the next century.[69] The J-10 program began in the late 1960s. After numerous technical difficulties, the Chinese asked the Israelis in the early 1980s to assist them with a new design based on the Israeli *Lavi* (U.S. F-16 derivative) fighter, and full-scale cooperation began officially by 1984. The J-10 project was launched on a full scale in 1987 and began receiving Russian assistance in the early 1990s.[70] The J-10 conducted its first test flight in March 1998, at least 2 years behind schedule, and is still undergoing testing and evaluation as more prototypes are produced. Predictions for deployment to an operational base range from 2003 to 2010, depending upon problems encountered in the test and evaluation phase. The J-10, which most likely will be armed with advanced beyond visual range (BVR), active radar (AR) air-to-air missiles (AAM) and may be air refuelable, is designed for counter-air and interdiction like the *Lavi* and F-16.[71]

One of the PLAAF's most significant events in the past 50 years was China's agreement with Russia in 1990 to purchase 48 SU-27 fighters and then to "produce" up to 200 of the aircraft in China over the next 10 to 15 years.[72] The SU-27s produced in China will be known as the J-11. The first batch of 26 aircraft arrived at the 3rd Air Division in Wuhu (Nanjing Military Region) in 1992, followed by a second group of 22 more at Suixi airfield (Guangzhou MR) in 1996. The first two domestically assembled Su-27s reportedly made their initial flights in December 1998.[73] According to the DoD report, the Su-27s are the only fighter aircraft in the PLAAF inventory with sufficient combat radius to allow extended operations beyond China's borders.[74] Like the J-10 program, there are differing opinions about the long-term viability of the Su-27/J-11 program.[75] While some analysts focus on the negative aspects of the program, such as Russia's restrictions on allowing China to produce the avionics and engines, other analysts note that China is moving ahead rapidly to develop its own engines and avionics for the aircraft and will not need the Russian systems.

Reports from Russia indicate that China is negotiating to purchase an additional 20 Su-27UBK two-seat trainer aircraft at a price of $35 million dollars each. The PLAAF originally received a total of four trainers along with 24 single-seat Su-27s for Wuhu and 22 single-seat Su-27s for Suixi.[76] Some analysts predict the PLAAF will purchase as many as 40-60 Su-27UBK to complement the full production run of approximately 200 J-11s. The PLAAF typically has four to six trainers per regiment. The Su-27UBK is fully combat capable and will likely be used to train pilots for both the Su-27SK and Su-30MKK. Delivery of the first trainers is expected within the next year, but the delivery schedule will probably match the production schedule of the J-11s at Shenyang, which will take the rest of the decade to complete.[77]

China is also negotiating with Russia to acquire and possibly co-produce the Su-30 multi-role aircraft that the PLAAF would use for deep strike/interdiction and air superiority. Although China is fully expected to receive some Su-30s, probably in 2001, there is still much confusion and speculation about the actual number and terms of the contract, including whether or not they will be co-produced in China. The figures have ranged from purchase of 20 to 72 aircraft and co-production of up to 250 aircraft.[78] There are also questions about whether the PLA Navy's Air Force will also receive the Su-30. Information from the 1999 Moscow air show indicates the Su-30MKK version being offered to China is significantly less capable than the Su-30MKI sold to India.[79]

Air-to-Air Missiles. The PLAAF has progressed further with its AAMs in the 1990s than any previous decade. Not only has the Air Force received advanced AAMs from Russia, but also pilots have had more opportunities to fire live missiles during training. According to the DoD report,

> The PLAAF currently has a number of advanced AAMs in its inventory, including the Russian-built AA-10a/Alamo and AA-11/Archer infrared (IR) AAM carried on the Su-27. China's J-7 is capable of carrying the PL-2A and PL-5B IR AAMs, as well as the all-aspect PL-8 IR AAMs, while its J-8IIs are capable of carrying the PL-2A, PL-5B, PL-8, and the beyond visual range (BVR) semi-active radar (SAR) PL-4 and PL-10 AAMs. By 2005, Beijing likely will have an active radar (AR) AAM in its inventory and could adapt it for use on a larger number of platforms. The PLAAF also is developing BVR AAMs for use aboard its J-10 fourth generation fighter aircraft.[80]

Bombers. The PLAAF received its first Soviet Il-28/*Beagles* in 1950 and TU-16/*Badgers* in 1959. Thereafter, China modified the Il-28 design and produced it in Harbin as the B-5. China also modified the TU-16 and produced it in Xian as the B-6.[81] The first B-5 and B-6 entered the PLAAF in 1967 and 1969, respectively. While the B-5 is no longer in production, the B-6 has been modified several times and is still produced primarily as a replacement for older airframes and for new missions, such as standoff cruise missiles and jamming. According to the DoD report,

> Today, China's bombers include the B-6 and the B-5. The B-5's slow speed and lack of standoff capability make this platform an extremely vulnerable target. The B-6 also is an aging aircraft, but it is being produced in several versions. One variant is designed to carry an antiship cruise missile (ASCM), while another is being developed to carry an air-launched cruise missile (ALCM). The B-5 is being phased out of the inventory,

but it is still used in training and would probably be employed along with the B-6 bomber during a military conflict.[82]

In addition, the Chinese have shown ECM variants of the B-5 and B-6. These aircraft were developed in the mid-1980s to provide standoff ECM support to their bomber fleet. While their effectiveness is unknown, intended victims would be enemy air defense early warning systems and their associated ground control intercept (GCI) structures that would impede interceptors from being vectored to an incoming bomber force.

Transports. One of the PLAAF's combat missions is to provide airlift in support of PLA operations. Until the mid-1990s, however, the PLAAF was unable to transport ground forces rapidly to distant parts of the country or sustain ground operations for extended periods due to antiquated aircraft and the lack of large-capacity aircraft. The PLAAF transport force is now capable of supporting the PLA at increased levels for a limited time and rapidly deploying to internal trouble spots. According to the DoD report,

> The PLAAF's current complement of large transport aircraft is limited to about a dozen Il-76/*Candids* and about 50 Y-8/*Cubs*, the remainder of the transport force consists of smaller aircraft like the An-24/*Coke*, An-26/*Curl*, and Y-5/*Colt*. Beijing can be expected to purchase a few additional Russian Il-76s or similarly sized foreign aircraft. The ongoing expansion of China's civil aircraft fleet will also allow the PLAAF to use the country's civil airlines to supplement its transport capability during crises.[83]

Use of civil aircraft is not new to the PLAAF. During the 1989 Tiananmen situation, civil aircraft were used to transport PLA troops to Beijing. In addition, military officials in New Delhi reported that the PLA used civil aircraft to ferry troops to Tibet during a recent exercise.[84] There are limits, however, to using civil aircraft to ferry troops into a hostile environment.

Airborne Early Warning. The PLAAF is also moving forward rapidly with its AEW program:

> In 1996, China signed a $250 million contract with Israeli Aircraft Industries (IAI) and Russia's Beriev to have IAI's subsidiary Elta install its Phalcon AEW radar in an Il-76 based Beriev A-50/Mainstay airframe. The aircraft will have a fixed radome housing a triangular phased array radar antenna to give the system 360-degree coverage, rather than using a rotating radome. The Il-76 will be equipped with electronic surveillance/electronic intelligence (ESM/ELINT) and communications intelligence (COMINT) systems. The first airframe was delivered to Israel in October 1999 and should arrive in China sometime in 2000. While the PLAAF reportedly has a requirement for 15-20 AEW aircraft, they currently have an option for only four airframes worth about $1 billion. Meanwhile, China has been negotiating with GEC-Marconi for an Argus radar system, which is also intended for an Il-76 but has not yet been mated to the airframe, and may still be flight tested against the Phalcon system before a final decision is made.[85]

The PLAAF has been searching for an AEW aircraft for about 15 years, including the unsuccessful modification of a Soviet-built TU-4 in the 1980s. Although the PLAAF should receive its first Il-76/Phalcon platform in the near future, some analysts have doubts whether the PLAAF will actually acquire all four aircraft. China has a history of trying to acquire one or two foreign systems and then having the aviation ministry try to copy them. There is a good possibility China might try to do the same with this AEW aircraft. Meanwhile, other analysts believe that the Il-76/Phalcon system will move forward rapidly.[86]

Intelligence Collection Aircraft. China maintains the most extensive signals intelligence (SIGINT) capabilities of all the countries in the Asia-Pacific region. The SIGINT systems include several dozen ground stations, half a dozen ships, truck-mounted systems, airborne systems, and a limited satellite collection capability. These capabilities have been significantly enhanced over the past decade, particularly with respect to electronic warfare (EW) operations. Aerial reconnaissance includes photographic and electronics intelligence (ELINT) activities. The principal Chinese airborne ELINT platform is the four-engine turboprop EY-8, an indigenous development of the Soviet An-12/Cub. The system is designed to detect, identify, analyze and locate land-based or shipborne radar emitters with a high probability of intercept. Israel has reportedly assisted development of China's airborne ELINT/EW capability, whereby several Chinese systems are derived from Israeli systems.[87]

Based on a *Jane's Defence Weekly* article, the PLAAF has also operationally deployed its first generation of locally developed long-range ELINT aircraft in early 1999. According to the report,

> Four modified TU-154 aircraft, fitted with radomes, have been deployed in the Nanjing MR opposite Taiwan, and at least eight more are on order. Although specifications of the aircraft are not known, some of the TU-154s may have limited airborne control and electronic warfare capabilities. Development of the aircraft was done at the PLAAF's Nanyuan airfield, located just south of Beijing.[88]

People's Liberation Army Naval Air Force. The PLA Naval air force was established in August 1950 as a separate administrative department within Navy Headquarters and as a separate PLA Navy operational branch. There are differing opinions concerning whether or not the Naval air force's modernization is keeping pace with the PLAAF. Although Naval air force has received the FB-7 and its bombers are equipped with air-to-surface cruise missiles, the PLAAF has received all of the Su-27s and most of the J-8II fighters so far. In addition, the PLAAF has begun flying over water, which heretofore was a Naval air force mission only. Naval air force's missions include protecting China's coastal airspace, providing air support for naval forces at sea, and conducting maritime search and rescue operations. The DoD report states,

> The PLANAF has only a limited maritime strike capability with some 150 non-standoff B-6s, A-5/*Fantans*, and B-5s. These aircraft would be only marginally effective against most modern navies. Some of the approximately 30 B-6Ds provide the PLANAF with a cruise missile ship interdiction strike capability, utilizing the C-601/Kraken ASCM. The standoff-capable FB-7 fighter-bomber, equipped with the C-801/ASCM, will become operational in the next 2 to 3 years. It likely will augment the B-6 and eventually replace some of the B-5s and A-5s in the PLANAF's inventory.[89]

According to interviews in Asia, the FB-7 is in the operational test and evaluation (OT&E) phase known as production finalization. The PLANAF already has 12 FB-7s in its inventory, although they may not be considered operational. Several of the aircraft overflew Beijing during the PRC's 50th anniversary on October 1, 1999. Various problems associated with the aircraft were revealed in a June 18, 1998, article in *People's Navy*, but these changes were minor modifications that took place during the OT&E phase:

> The PLANAF had to improve the FB-7 after receiving them. When the aircraft first arrived, it was still being tested, manufactured, and improved. The transceivers easily burned out, and the fuel tanks had to be replaced periodically. As a result, the regiment added a protector to the transceiver, and devised an easier way to replace the fuel tank. More innovations followed, including checking the fueling procedure without starting the engine; checking the fueling valve without increasing pressure; a faster way to hang, connect, and test a model of a missile; and a device to prevent engine cough caused by exhaust from missile launching that led to engine shutdown in mid-air. The regiment also found ways to protect the new equipment against heat, humidity, corrosion, rat infestation, and typhoons.[90]

The PLANAF has also secured the purchase of their own AEW aircraft, the SkyMaster AEW system from the United Kingdom. According to *Defense News* they have purchased up to eight AEW systems for installation on the Y-8. The *SkyMaster* includes both AEW and maritime patrol functions, which fits well with PLANAF plans. While it would be of marginal use against a dedicated fighter attack, it is well suited to direct intercept of long-range maritime patrol aircraft.[91]

Air Defense Troops.

During the 1990s, the PLAAF has upgraded its air defense (nonaircraft) capabilities. This involves three of the PLAAF's branches: SAM, AAA, and radar troops. While the PLAAF has modernized its SAMs significantly in terms of quality, if not quantity, the overall air defense network is quite weak. According to the DoD assessment,

> Beijing is expending tremendous effort establishing an Integrated Air Defense System (IADS) at both the strategic and tactical levels. China's air defense technology currently lags behind western standards and its current IADS capability lacks many crucial components. Beijing probably could establish a fully operational national IADS within the next twenty years, but clearly not by 2005. China has a rudimentary tactical IADS capability in the form of its mobile Tactical Air Defense System (TADS).[92]

The PLA readily admits that it has problems with its air defense network. For example, a July 14, 1998, article in *Jiefangjun Bao* provided insight into this problem. The article stated that,

> Air battles in the 21st century will be a form of combat which shows information warfare at its highest. In order to seize superiority in information, the attacking side will adopt electronic suppression, electronic deception, and route selection to prevent the defensive side from reacting on time. Stealth planes, in particular, will greatly reduce the detection range. The PLAAF will have to advance its air defense early warning equipment in order to be able to detect invading targets early, extend the early warning range, and seize the necessary time and space for fighting air defense battles.[93]

The article further stated that,

As precision guidance technology develops, the launch distance of different types of air-to-ground missiles and aerial guided bombs will increase. These advancements will result in combat aircraft changing their methods of launching surprise attacks against ground targets. Instead of overhead bombing, the aircraft can now launch munitions from outside the enemy's range of fire or even from outside the air defense area. If the PLAAF continues to use the method of air defense at key points or key areas, attacking aircraft will have already completed their bombing and safely returned home before the PLAAF's air defense system could even react. Therefore, the PLAAF must enlarge its effective air defense area and shoot down enemy aircraft before they can even launch their air-to-ground weapons. The PLAAF must persist in a dynamic air defense system by moving forces rapidly and concentrating firepower in order to enlarge the effective air defense area. The PLA must also organize all of its air defense forces under centralized and unified operational command.[94]

Given the PLAAF's acknowledged weakness in these areas, Commander Liu has stated the Air Force's SAM units are engaged in an evolution from single-type air defense missiles and equipment to new integrated air defense systems, combining high, medium, and low-altitude, long and short-range missiles.[95]

Since the early 1990s, the PLAAF has purchased two regiments of S-300 (SA-10) and one regiment of Tor-M1 (SA-15) SAMs from Russia. During late 1998, Beijing was reportedly negotiating with Moscow to acquire two additional S-300 regiments and one Tor-M1 regiment.[96] Although most of the S-300s are currently based near Beijing, some of them were reportedly deployed near Fuzhou during the 1996 Taiwan Strait exercise for exercise play and as a precautionary measure.[97]

China has also developed the FT-2000 mobile SAM. This SAM is based on the S-300 and is designed to engage radiating aircraft, such as airborne jammers and airborne warning and control system (AWACS) aircraft. China first tested the FT-2000 in September 1998, then conducted the first field trials during a series of exercises in 1999.[98]

Air Activity Over the Taiwan Strait: A Case Study.

On July 9, 1999, President Lee Teng-hui in Taiwan declared that relations between Taiwan and mainland China should be conducted on a "state-to-state" basis. Shortly thereafter, the PLAAF increased the number of sorties over the Taiwan Strait, including crossing over the centerline of the Strait several times.[99] This issue generated a lot of interest and speculation about the possibility of an aircraft incident occurring, either by accident or on purpose by either side, that could start a war across the Strait.

On August 10, a Taiwan Ministry of National Defense (MND) spokesperson stated that:

> eight J-8 fighters have deployed to Fuzhou airbase and are conducting flights over the Taiwan Strait. There are about 150 aircraft permanently stationed in Fujian's coastal areas. In addition, Su-27 fighters stationed in Guangdong Province have begun training activities in areas close to the center line.[100]

Although not stated, the Su-27s were probably flying direct missions to and from the Strait from their home base at Suixi. The Su-27s stationed at Wuhu were probably also involved.

Although the number of PLAAF sorties increased following President Lee's statement, this was not really new activity. The PLAAF was actively involved during the PLA's large-scale exercises opposite Taiwan during March 1996. According to available open source material,

> The exercise included 12,000 Air Force and 3,000 naval air force servicemen. More than 280 aircraft deployed to the exercise area and conducted 680 sorties, including eighty-two transport sorties. Over 800 combat aircraft were within a combat readiness of 550 miles or were on the alert.[101]

Another report stated the PLA deployed fewer than 100 additional aircraft to the thirteen Fujian airfields from other bases, raising the total to only 226 aircraft. Based on a briefing by the U.S. Office of Naval Intelligence, the PLA¹⁰² conducted a total of 1,755 sorties during the exercise. Further press reporting stated that the PLAAF deployed aircraft from its second and third-line airfields to first-line airfields, where they conducted their exercise activity. It took about 3.5 hours for the PLAAF fighters to prepare for takeoff, compared to the 10 hours they had needed previously. In addition, the PLAAF demonstrated rapid aircraft sortie regeneration of 40 minutes, which was considerably quicker than in the past.¹⁰³

In November 1998, the commander of a Taiwan Air Force *Mirage*-2000 fighter group stated,

> Following the air battles that took place over the Taiwan Strait in 1958, our fighters have kept a distance of 35 miles from the mainland's coast, while the Chinese Communist fighters usually carry out their duties close to their own coast line. If two Communist jet fighters took off, the Taiwan Air Force would dispatch four planes to watch them. Maintaining a tacit agreement on an invisible centerline of the Strait, neither side has conducted any provocative flights. In the past, there was a tacit agreement that "we leave when you come, and we come when you leave." Recently, however, the Chinese Communist fighters have conducted frequent intentional flights across the central line of the Taiwan Strait. Taiwan Air Force F-16 and Mirage 2000 aircraft could detect Chinese Communist fighters on their radar screens. The Communist aircraft were probably attempting to collect information about the training of our new-generation fighters.¹⁰⁴

There are several explanations for the increased activity, in addition to reacting to President Lee's statement. Since its founding in 1949, the PLAAF has consistently tried to carry out its mission of controlling China's airspace. During the early 1950s, the PLAAF was a fledgling service and focused all of its efforts in the northeast and on the Korean War. During that period, Taiwan's Air Force controlled the skies over southeast China as far north as Shanghai. Therefore, besides defending the border with Korea, the PLAAF's immediate task was to provide air defense for the major cities, such as Shanghai, Beijing, and Tianjin. It was not until after the 1958 Taiwan Strait crisis that the PLAAF had a permanent presence in the provinces opposite Taiwan. Even then, Taiwan continued to fly U-2 reconnaissance flights over the mainland until the late 1960s. Since 1958, there have not been any serious air engagements over the Strait.

According to a Taiwan Ministry of National Defense (MND) spokesman in August 1999, "Following Beijing's 1996 military exercise, Communist Chinese aircraft have been flying more often in the Taiwan Strait."[105] Although the media began focusing on the PLAAF's increased air activity after President Lee's statement, a knowledgeable source in Taiwan indicated that the PLAAF began increasing its flight activity near the centerline in June 1998.[106]

There are several probable reasons why the PLAAF has stepped up its flight activity over the Strait since 1996 and near the centerline in 1998. First, the CMC wanted to challenge the fact that Taiwan's Air Force has basically owned the entire airspace over the Taiwan Strait since the 1950s and the PLAAF could not effectively challenge this control. Taiwan's fighter and reconnaissance aircraft have routinely crossed the centerline, and transport aircraft ferry supplies to troops on Jinmen and Mazu. The PLAAF had little choice but to have a tacit agreement on the centerline, so that it could at least have the facade of some control over the Strait. Even so, the mainland officially refutes the idea of a centerline, stating that Taiwan is a part of China, so there cannot be a centerline over a body of water that belongs to China.[107]

The PLAAF also probably felt compelled to react to significant changes in Taiwan's Air Force or lose the opportunity to have any presence at all over the Strait. These changes included Taiwan's Air Force commissioning its first indigenous defensive fighter (IDF) squadron in December 1994, and the first French *Mirage* 2000-5 and

227

U.S. F-16 squadrons in October 1997.[108] Liu Shunyao, who became the new PLAAF commander in December 1996, began emphasizing the PLAAF's offensive capabilities when Taiwan received the first *Mirages* and F-16s in April 1997.

During July and August 1999, only 12 PLAAF aircraft were airborne at any time, not all of which were over the Strait, and the PLAAF flew only about 30 total sorties per day.[109] However, these sorties were a significant change for both air forces. Over 1,000 civil air flights fly through Taiwan's airspace daily. Most of the airspace immediately north and south of Taiwan, flying to/from Taipei and Kaohsiung, is dedicated to civil air routes, which means that Taiwan's Air Force is forced to conduct its training in the Strait and to the east of the island. These sorties, and any future such activity, force Taiwan's Air Force to divert aircraft from their regular training regimen to conduct patrol duties (punching holes in the sky) in the Strait. Although the PLAAF did not fly that many sorties in the Strait, Beijing sent a clear message that the PLAAF could fly in the Strait if it wanted to.

The PLAAF's Force Opposite Taiwan. Much of the PLA's modernization efforts over the past decade have focused on Taiwan. While the PLA has had a sizeable force in the Nanjing Military Region opposite Taiwan since the 1958 Strait crisis, the biggest change during the 1990s involved improving the quality of the PLAAF's equipment. According to the *Republic of China: 1998 National Defense Report*,

> The PLAAF can station 1,200 combat aircraft and maneuver 59 air transports to carry two airborne regiments for operational missions out of the thirteen military-civilian airports within 270 miles of Taiwan proper. The PLAAF stations and deploys its aircraft into three areas based on their distance from Taiwan: first line (270 miles), second line (550 miles), and third line (beyond 550 miles). At present, the PLAAF has 1,300 aircraft stationed on the air bases within 550 miles from Taiwan, of which 600 have a radius of operation over Taiwan proper. The Su-27s, with a combat radius of just less than 900 miles, can also

cover Taiwan from their bases at Wuhu, Anhui Province, and Suixi, Guangdong Province.[110]

Under current circumstances, once the PLAAF has deployed several hundred aircraft near Taiwan, it must then coordinate an air attack. Although the PLAAF has been practicing this, it is not there yet. The chief difficulty is simply exercising strict ground control intercept (GCI) over large numbers of aircraft simultaneously. According to a U.S. Office of Naval Intelligence briefing, during the 1996 exercises, PLAAF and PLA Naval Air Force fighters and bombers engaged in simulated and live bombing, fighter escort for bombers, air-to-air combat training, and other routine training. The PLAAF's airborne forces were also involved in airdrop activity.[111]

The key to any conflict for the PLAAF is sustained combat, and the PLAAF has not yet demonstrated the capability to conduct sustained, high intensity operations. The PLAAF does not have any real-world experience in planning and executing the kind of high intensity air campaign that has proven so successful in recent U.S. and NATO operations. For example, during the early stages of the conflict in Kosovo, allied air forces deployed approximately 400 aircraft to the area. On the third day of operations alone, allied aircraft flew 249 sorties in one day. By the end of the conflict, the number of U.S. and NATO combat aircraft participating in strike delivery rose from 214 to 590 aircraft. According to Pentagon information, 23,000 bombs and missiles were used. The maximum intensity of operations of Allied Air Forces was reached on May 21 when 1,000 sorties were flown, 800 of which were combat missions. These figures demonstrate the capability needed to ramp up and maintain high intensity operations, orchestrate operations through a unified daily air tasking order (ATO), and the need to sustain intense air operations when faced with a determined adversary.[112]

Interviews with Taiwan and Japanese officials in early 1999 indicate that the PLAAF has been rotating J-7 and J-8 aircraft from designated rapid reaction units in and out of Fujian Province on 6-month rotations since the 1996 crisis, in order to conduct area familiarization training. Furthermore, aircraft from Fujian Province have been deploying to and from airbases outside Fujian during the same day, providing them with the ability to move out of the area rapidly if the need arose. In addition, all PLAAF pilots within the Nanjing Military Region have been conducting live air-to-air missile firing over water since 1996. These aircraft have probably been designated as part of the PLAAF's rapid-reaction force for action against Taiwan.

One would expect these aircraft to be deployed to the area *en masse* if the PLAAF were going to attack Taiwan. For example, during the 1979 border conflict with Vietnam, the PLAAF deployed almost 20,000 aviation, SAM and AAA troops and 700 aircraft to the border area opposite Vietnam over a 45-day period leading up to the start of the conflict. Yet none of these aircraft engaged in any activity across the border so as not to escalate the conflict based on the tacit rules-of-engagement (ROE) for both sides at the time.[113]

There is always the possibility of an accidental shoot down occurring, but there is a low probability, unless it is pre-planned. The reason for this is that each side has established unilateral ROEs. For example, Taiwan's Air Force is under the guidance to "neither avoid the Communist planes nor provoke them. The order to attack an intruder must come from Taiwan's Chief of the General Staff."[114] On the mainland side, the PLAAF probably has specific ROEs as well. For example, during the 1958 Strait Crisis, the PRC's Central Military Commission established the following three ROEs: (1) the Air Force could not enter the high seas to conduct operations; (2) if the Nationalist Air Force did not bomb the mainland, the PLAAF could not bomb Quemoy and Matsu; and, (3) the Air Force was not allowed to attack the U.S. military, but could defend against any U.S. aircraft entering Chinese territory.[115]

The current aerial cat and mouse game over the Strait actually helps each side prepare for potential conflict, while at the same time helping to avoid conflict. Each side has the opportunity to determine how long it takes for the other side to scramble its aircraft upon detection of an intruder, to reach a specific spot over the Strait, and to determine what altitude and speed these aircraft conduct their missions. Each side can also test the limits by intruding into the other's airspace to see what type of a reaction they get. They also have the opportunity to employ their air intercept radar to detect, track, monitor and shadow opposing aircraft, significantly enhancing their confidence in their defensive capability. Concurrently, these defensive reactions against the probes by intruder aircraft demonstrate and reinforce a vigilant air defense posture to the opposing force.

These types of exercises also afford each side the opportunity to coordinate their flight activity with the ground-based air defense systems, including their radar, antiaircraft artillery, and surface-to-air missile units. For example, one of the biggest problems the PLAAF, and the PLA as a whole, will face if China engages in a conflict with Taiwan is coordination between aircraft and ground-based air defense forces. Ever since the Korean War, the PLA Air Force and Army have had to agree on some basic ROEs, so that the Army's and PLAAF's AAA and SAMs do not shoot down the PLAAF's aircraft. This is even further complicated if they must also coordinate with the Navy.

Today, the primary means of coordination between PLAAF aircraft and ground-based air defense units is airspace differentiation. The PLA's air defense system relies heavily on airspace control measures and procedural control (essentially flying air corridors, adhering to strict altitudes and in-area time limits, or speed controls). Although most, if not all of the PLAAF's aircraft have some type of an identification friend or foe (IFF) system, the best method for coordinating aircraft movement with the ground forces or a ground controller is through the use of a secure and reliable electronic identification (EID), which utilizes a reliable IFF or other means. Moreover, it is not clear whether the older J-6 and J-7 aircraft have adequate navigation equipment (inertial navigation system/INS is desirable because it is an on-board system) to maintain a reasonable adherence to air corridors during flights over water. The FB-7 has an INS and doppler navigation, and most likely has TACAN. The PLAAF recognizes these weaknesses and recent PLA news reports have highlighted the PLAAF's efforts to conduct training over water for their pilots.

In the situation over the Taiwan Strait, just because an aircraft "locks on" to another aircraft with its radar, does not necessarily mean that it is going to fire a missile. Most pilots or ground controllers do not have the authority to fire a missile unless told to do so by much higher authority. In this case, the intruder will most likely have already peeled off and returned to his airspace. This time provides the opportunity for each side to take a deep breath and cool down. In the current situation, I would say that the probability of an accidental firing is fairly minimal.

If war does break out, there will be more than just aircraft involved. The mainland will most likely begin the attack on Taiwan with ballistic missiles in order to degrade Taiwan's command and control capability, as well as to try to destroy aircraft on the ground before they can take off. Although PLA writers have advocated a quick strike on Taiwan to achieve victory within 2 weeks so that the United States cannot come to Taiwan's aid, this is easier said than done. There will be definite political, as well as military signals, coming from the mainland that should provide enough time for all parties, including the United States, to try to achieve some type of political settlement and/or pre-position military resources in the region before war breaks out.

Conclusion.

There is no simple description of the PLAAF's operational capabilities. The PLAAF is in the process of modernizing, but it still has a long way to go. Parts of the PLAAF are clearly obsolete, yet other parts have the most modern, sophisticated aircraft, SAMs, and software. The key to the PLAAF's modernization is integrating all of the different components, including its branches and new and old weapon systems, into a single operational unit within the PLAAF and with the rest of the PLA as a whole. As one China watcher recently stated, "It is not just a matter of the glass being half empty or half full, because the glass is getting bigger."

During discussions of the PLAAF, a former U.S. Air Force Chief of Staff identified the following questions that he would want to know about China's airpower:

- What is their battle doctrine?

- Is their organizational direction defensive or offensive?

- Which classic air missions do they cover effectively—air defense, counter-air, conventional strike?

- What is their basic employment concept for these air missions—manned aircraft, ground or air launched missiles?

- What kind of support packages do they have—defense suppression, air escort, tanker support?

- What kind of power projection capability does their equipment represent—range, payload?

- How modern is the capability—precision guided weapons, battlespace surveillance?

- What is their equipment and personnel readiness status?

- What size, duration, and location of operations can they sustain with logistics support?

As discussed in this chapter, the PLAAF clearly comes up lacking in most of these categories. However, during interviews in Japan, Taiwan, Southeast Asia, and India, defense officials repeatedly said that the United States cannot just look at the PLAAF through Western eyes. Everyone in these countries agreed that the PLAAF is not necessarily capable of sustained combat today, but it is moving in the direction of a more modern offensive force over the next 15 to 20 years. Each of these countries is looking carefully at China's intentions and taking the long-term view of China's potential capabilities.

At the same time, however, it is my opinion based on interviews in the region that most of the countries surrounding China do not have a thorough understanding of the PLAAF. For example, other than Taiwan and Japan, there are very few assets or analysts dedicated to just the PLAAF. The majority of the countries in Asia have only one military attaché in Beijing, most of whom do not speak Chinese, and who concentrate primarily on the macro-level PLA strategic issues, rather than on operational capabilities. In addition, most of the countries in Asia focus only on the PLAAF forces in their immediate border region, rather than on the PLAAF as a whole. They also tend to focus on hardware issues only (i.e., how many Su-27s does China have), rather than on the software issues.

The PLAAF is still primarily a defensive force, but, through the acquisition of systems with longer ranges and more lethal bombs and missiles, it is moving gradually toward having an offensive capability. Although the PLAAF is acquiring some systems to support the air defense mission, such as refueling, ECM, and airborne early warning platforms, it will be several years before these systems can be fully integrated into the force, and even then, only in limited numbers. The PLAAF still lacks precision guided munitions and a battlespace surveillance capability, but China is working on these capabilities. The PLAAF is beginning to train its forces in mobile operations, but is still hindered by institutional, organizational, and equipment limitations that hinder mobile operations. Furthermore, the PLAAF is not yet capable of round-the-clock or sustained operations.

There is little doubt that what the PLAAF has done over the past decade is impressive. The Air Force has acquired limited numbers of Russia's most sophisticated weapon systems, but still must rely on Russia for long-term logistics support. The PLAAF will continue to rely on imported weapons systems to modernize its force, but is moving toward developing and producing better domestic engines, avionics, and missiles. Air combat training has become more realistic, including more live air-to-air missile launch training, but the pilots still lack sufficient flying hours due to engine and airframe limitations. While the PLAAF's Blue Force aggressor program has obviously been successful, it has pointed out even more starkly the limitations of aircraft like the J-6, J-7, and J-8.

With the acquisition of several Russian Il-76 transports, the PLAAF's 15th Airborne Army has been able to train in more locations around China, but the focus is still on use for domestic situations rather than situations outside China's borders.

Although Commander Liu Shunyao states that the PLAAF has the capability to support ground and naval operations, the Air Force's ability to do this is questionable. The PLA as a whole is just beginning to address joint operations and logistics. Very little information is available in open source material about PLAAF support for the PLAN, which would entail joint exercises and some type of joint command and control structure. As for supporting the ground forces, the PLAAF has consistently stated that it cannot provide direct support for the ground forces. Nothing has occurred over the past decade to change this philosophy. Su-30 will not replace the A-5. While the Su-30 will replace the B-5 in the interdiction and deep strike mission, there are not ready replacements for the A-5, whose mission is battlefield interdiction and ground attack.

While the PLAAF can look proudly at its accomplishments in equipment acquisition and training over the past decade, the bulk of the aviation force still consists of aging J-6, J-7, and J-8 interceptors, only some of which have an all-weather capability. China's aviation industry is working toward designing and producing a modern aircraft to meet the PLAAF's needs in 15 years, but will still need foreign assistance from Russia and Israel along the way. Therefore, the PLAAF will still be relying on Russian support for the near term. Once China's next generation of domestically produced aircraft, such as the J-10, improved FB-7, and XXJ, are operationally deployed, we will have a better understanding of how far China's aviation industry has come.

The PLAAF will continue to analyze its needs and implement changes as it can, but there will also continue to be resource and system limitations that will hinder the PLAAF's overall modernization drive and operational capabilities. China will eventually have to face the fact that there are limitations on how many foreign aircraft it can buy. If the PLAAF is going to replace its current fleet of aircraft, it will have to rely on China's aviation industry to provide much of the necessary components or the Air Force will not be able to afford it. The PLAAF simply cannot afford to continue purchasing Russian aircraft at $35-40 million apiece, plus rely on the Russian supply line. Finally, the PLAAF's leadership must also accept aircraft losses as part of a more vigorous, realistic training program.

In closing, a clear distinction must be made between the PLAAF's capabilities and intentions. Even if the PLAAF is not the most modern force, the PLAAF's leaders will salute smartly and use every means available to achieve victory if the CMC tasks them to go into battle.

CHAPTER 6 - ENDNOTES

1. *"Jiefangjun Bao* Calls For Stronger Air Force," Hong Kong *AFP*, April 7, 1996, *Foreign Broadcast Information Service-China (FBIS-CHI)*-96-068, April 7, 1996.

2. "China: PRC Air Force Plans Weapons Advances," *Xinhua*, April 14, 1997, *FBIS-CHI*-97-104, April 14, 1997.

3. "Deputy Commander Says Air Force Improves Combat Readiness," *Xinhua*, January 20, 1999, *FBIS-CHI-99-020*, January 20, 1999. These comments were made during a work conference on Air Force training.

4. Oliver Chou, "President calls for hi-tech push by Air Force," *South China Morning Post*, March 3, 1999.

5. While researching this paper, the author reviewed numerous articles written about the PLAAF from within and outside of China, hosted a series of meetings on airpower in Asia, and visited Taiwan, Japan, China, Singapore, Malaysia, Vietnam, and India, to conduct interviews about the PLAAF.

6. *Dangdai Zhongguo Kongjun (China Today: Air Force)*, Beijing: China Social Sciences Press, 1989. PLAAF Headquarters Education and Research Office, *Kongjun Shi (History of the Air Force)*, Beijing: PLA Press, November 1989. The PLAAF describes these air battles as occurring during involvement in "liberating Tibet," the "War to Resist America and Aid Korea," numerous engagements with Nationalist and U.S. aircraft over the Taiwan Strait, the "War to Aid Vietnam and Resist U.S. Aggression," and the 1979 "self-defensive counterattack" against Vietnam. During the Korean War, the Chinese Air Force reportedly shot down 330 American planes and damaged 95. During the 1958 Taiwan Strait crisis, PLAAF data indicates the air force and naval air force aircraft shot down 14 and damaged 9 aircraft. During the Vietnam War, the PLAAF states that its AAA units were involved in 558 battles, shooting down 597 U.S. aircraft and damaging 479. Like the figures cited during the Korean War and Taiwan Strait Crisis, they may not agree with Western figures. In addition to aircraft that its AAA shot down over Vietnam, the PLAAF history describes in detail how its pilots shot down three U.S. Air Force aircraft (one F-4B and two A-4Bs) and two Navy (one A-3B and one A-6) aircraft and 17 unmanned reconnaissance drones over or near Chinese territory. The official PLA Naval Air Force history also identifies two U.S. drones and four aircraft (F-4B, F-104C, F-4C, and A-1) that naval air force pilots shot down over Hainan Island. No air combat took place during the 1979 border conflict

with Vietnam. The August 1998 issue of *Xiandai Junshi* is devoted to the 70th anniversary of the PLA and carries additional information about aircraft the PLAAF shot down during the 1950s and 1960s.

7. Over the past 15 years, China has improved its marketing techniques for aviation equipment and has hosted an annual airshow at Zhuhai. One of the problems associated with this, however, is that some analysts assume that the aviation ministry and PLAAF have incorporated or are going to incorporate the systems being marketed by Chinese research institutes and companies into PLAAF aircraft. This is not necessarily the case. Many of the systems are marketed for sale abroad only and are not for domestic use. Furthermore, some analysts assume that just because Russian salesmen appear at Zhuhai and say the Chinese are interested in their systems, that China is going to purchase them.

8. For an overview of the PLAAF's logistics and maintenance capabilities, see Kenneth W. Allen, "PLA Air Force Logistics and Maintenance: What Has Changed?," in *The People's Liberation Army in the Information Age*, James C. Mulvenon and Richard H. Yang, eds., RAND, 1999, pp. 79-97.

9. Sun Maoqing, "PLA Commander on Modernizing Air Force," *Liaowang, FBIS*, No. 15, April 14, 1997, pp. 20-21. Hua Chun, Chang Tun-Hua, and Kuo Kai, "Air Force Trains Crack Units Troops for Offensive, Defensive Operations," "Interviewing Lieutenant General Liu Shunyao, PLA Air Force Commander," Hong Kong *Ming Pao*, August 2, 1997, *FBIS-CHI-97-226*, August 2, 1997. Speech by Liu Shunyao, "Comprehensively Push Forward PLA Modernization Building," *Jiefangjun Bao, FBIS*, December 24, 1998.

10. *Ibid.* The whole issue of the PLA's rapid reaction (*kuaisu fanying*), or fist (*quantou*), units is rather murky, especially as they pertain to the Air Force. According to Mel Gurtov and Byong-Mao Hwang, *China's Security: The New Roles of the Military*, Boulder, CO: Lynne Rienner Publishers, Inc., 1998, pp. 107-108, the PLA has RRFs and rapid deployment forces, (RDFs), which comprise around 11 percent of the total ground forces. The RRF is composed of airborne units and light infantry units. They are equipped with light weapons and depend entirely upon air transportation, including transports and helicopters. They are expected to react rapidly to border disputes, minority rebellions, and political violence; the airborne forces were utilized in Tiananmen during June 1989. The RDFs are equipped with heavy weapons and will quickly deploy to deal with any contingencies, only after its RRF reacts. Gurtov and Hwang do not mention anything about aviation RRF units.

11. The PLAAF's definition of flying in "weather conditions" is divided into "three weather conditions", i.e., day and night visual flight rules (VFR), and day instrument flight rules (IFR)), and "all-weather" or "four weather conditions" which adds night IFR flights. Although this particular reference does not mean being able to fly in poor weather conditions, some reference to flying in weather conditions does mean exactly that. The exact meaning is usually clear. The activity in 1998-1999 compares to information from 1991, where 70 percent of PLAAF pilots at that time were qualified as all-weather pilots. Training in 1991 also included low-altitude flying, minimum-altitude flying, air combat, high-altitude air acrobatics, and transregional movement. Eighty percent of the PLAAF's air combat regiments had reportedly taken part in multi-service and multi-arm tactical exercises in 1991. Kenneth W. Allen, Glenn Krumel, and Jonathan D. Pollack, *China's Air Force Enters the 21st Century*, Santa Monica: RAND, 1995, p. 130.

12. Huang Qiang, "Will China's Aviation Industry Be Able to Get Out of the Doldrums Soon?," *Keji Ribao (Science and Technology Daily)*, July 8, 1999, *FBIS-CHI*-1999-0811, July 8, 1999.

13. The PLAAF has about 20 research institutes. Eight are subordinate to the Scientific Research Department within the Headquarters Department; each has about 180-200 personnel. In addition, the Logistics Department also has at least eight subordinate institutes. Other departments also have specialized research institutes. Kenneth W. Allen, *People's Republic of China, People's Liberation Army Air Force*, Washington, DC: Defense Intelligence Agency, May 1991, Section 17.

14. For a detailed analysis of this revised national military strategy, see Nan Li, "The PLA's Evolving Warfighting Doctrine, Strategy, and Tactics, 1985-95: A Chinese View," and Paul H.B. Godwin, "From Continent to Periphery: PLA Doctrine, Strategy, and Capabilities Toward 2000," both in David S. Shambaugh and Richard H. Yang, eds., *China's Military in Transition*, Oxford: Clarendon Press, 1997, pp. 284-312. See also James Harris, *et. al.*, "Chinese Defense Policy and Military Strategy in the 1990s," *China's Economic Dilemmas in the 1990s: The Problems of Reforms, Modernization, and Interdependence*, Washington DC: Joint Economic Committee, Congress of the United States, April 1991, pp. 648-649. The concept of active defense means taking tactically offensive action within a basically defensive strategy. The defending forces undertake offensive operations to wear down the adversary while he is strategically on the offensive and attacking. It is the opposite of passive defense, which means the defending forces simply resist without attempting to weaken the adversary as he prepares to attack or is actually on the offensive. The active defense

strategy consists of three phases: strategic defense, strategic stalemate, and strategic counterattack.

15. Paul H.B. Godwin, "Compensating for Deficiencies: Doctrinal Evolution in the Chinese People's Liberation Army," presented at the 1999 CAPS-RAND PLA Conference, pp. 7, 43.

16. Hua Renjie, Cao Yifeng, and Chen Huixiu, editors, *Kongjun Xueshu Sixiang Shi*, Jiefangjun Publishers, Beijing, 1991, pp. 312-331.

17. Sun Maoqing, "Make Efforts To Build Modernized People's Air Force: Interview With Air Force Commander Lieutenant General Liu Shunyao," Beijing *Liaowang*, April 14, 1997, No. 15, pp. 20-21.

18. "Junshi Jingji Yanjiu (Military Economics Studies)," No. 8, 1995.

19. "PLA Airborne Brigades Become Divisions," *Jane's Defence Weekly*, Vol. 20, No. 14, October 2, 1993, p. 12.

20. "The Security Situation in the Taiwan Strait," Report submitted by Secretary of Defense William Cohen to the U.S. Senate as directed by the FY99 Appropriations Bill, February 17, 1999.

21. Sun Maoqing, "Make Efforts To Build Modernized People's Air Force: Interview With Air Force Commander Lieutenant General Liu Shunyao," Beijing *Liaowang*, *FBIS*, April 14, 1997, No. 15, pp. 20-21.

22. "China Employs Hi-Tech Equipment in Training To Improve Air Force Fighting Capacity," Hong Kong *Sing Tao Jih Pao*, July 18, 1997. The PLAAF has written several articles emphasizing that airborne forces have been training with a new type of parachute.

23. "Airborne Units Conduct Training Exercise in Tibet," Beijing Xinhua, *FBIS*, June 29, 1999.

24. "Chinese Exercise Strait 961: 8-25 March 1996," briefing presented by the U.S. Office of Naval Intelligence at a conference on the PRC's military modernization sponsored by the Alexis de Tocqueville Institute, March 11, 1997. Although Commander Liu has touted the airborne army's all-weather capabilities, they are limited by the planes' capabilities to get them there.

25. Interviews during April-June with air force officers in Asia. Although interviews during April through June 1999 indicate that the PLAAF has established certain aviation units as RRF units, particularly those with J-7, J-8, and Su-27 fighters, no open source

material was found to corroborate their analysis. Even so, the PLAAF probably has organized certain aviation units as RRF units.

26. Nan Li, "The PLA's Evolving Warfighting Doctrine, Strategy, and Tactics, 1985-95: A Chinese View," in David S. Shambaugh and Richard H. Yang, eds., *China's Military in Transition*, Oxford: Clarendon Press, 1997.

27.*Shijie Junshi Nianjian* (World Military Yearbook), was published by PLA Press, Beijing, in 1985, 1987, 1989, 1990, 1991, and 1992. Each yearbook provided progressively greater detail about the PLA's organizational structure.

28. "Air Force, Navy To Enforce Fishing Ban in South China Sea," Beijing *Xinhua*, May 7, 1998.

29. The term *kongjun* means the Air Force as a whole, but it also means Headquarters Air Force. Subordinate elements in the chain-of-command include the seven military region air forces/MRAFs, (*junqu kongjun*), air corps (*jun/kongjun jun*), command posts (*zhihuisuo*), and operational units (*budui*), which include air divisions (*shi/kongiun shi/hangkong bing shi*), brigades (*lu*), regiments (*tuan*), groups (*dadui*), squadrons (*zhongdui*), battalions (*ying*), companies (*lian*), platoons (*pai*), and squads (*ban*). The Air Force, which is a service (*junzhong*), is composed of five branches (*bingzhong*).

30. The four first level departments are headquarters (*silingbu/kongsi*), political (*zhengzhibu/kongzheng*), logistics (*houqinbu/konghou*), and armament (*zhuangbeibu/kongzhuang*). When the PLAAF was founded in November 1949, an Air Force Engineering Department (*kongjun gongchengbu*) was established to manage aircraft maintenance, and the PLAAF's Engineering College in Xian was established in September 1964. Maintenance for all other PLAAF equipment, vehicles, SAMs, AAA, etc., was the responsibility of the Logistics Department. In September 1969, the Engineering Department was abolished, and its Field Maintenance Department (*waichangbu*) was subordinated to the Headquarters Department as the Maintenance Department (*jiwubu*). On May 1, 1976, the Aeronautical Engineering Department (*hangkong gongchengbu/konggong*) was established as a first level department. In October 1998, the Aeronautical Engineering Department was reorganized to became the PLAAF Armament Department (*zhuangbeibu/kongzhuangbu*) in order to be in alignment with the PLA's new fourth general department—the General Armament Department (*zongzhuangbeibu*).

31. In order to exercise the Party's absolute leadership over the military, a Party Committee and Standing Committee is established at each regiment (and equivalent) and above. The political commissar is normally the secretary of the Party Committee at these levels. Grassroots Party Committees are established at each battalion and equivalent. Grassroots Party Branches are set up at the company level. Besides the Party Committee, political organizations have been established in each regiment and higher, and their equivalent. The General Political Department (GPD) is the highest leading body for political work in the PLA. While divisions and brigades have a Political Department, regiments have a Political Division.

32. The official term for the PLAAF's AAA troops is *gaoshepao bing/gaopaobing*, and for the SAM troops *dikong daodan bing/didao bing/daodan bing*. However, the PLAAF occasionally refers to its AAA troops as first artillery (*yipao*), and SAM troops as second artillery (*erpao*), which is often confused with China's Second Artillery Corps (*erpao*).

33. Unless otherwise specified, this section describing the historical organizations of the PLAAF come from interviews with PLAAF officials plus Kenneth W. Allen, *People's Republic of China, People's Liberation Army Air Force*, Washington, DC: Defense Intelligence Agency, May 1991, Section 13.

34. Interview with PLAAF officials.

35. "The Security Situation in the Taiwan Strait," Report submitted by Secretary of Defense William Cohen to the U.S. Senate as directed by the FY99 Appropriations Bill, February 17, 1999, p. 6. Discussions with PLAAF officials indicate that this number (32 divisions) is too low, but they did not specify the total number. In addition, the number of 4500 aircraft is probably too high. According to a March 1997 Hong Kong report, U.S. reconnaissance satellites discovered in June 1993 that China had gathered over 1,000 combat aircraft at an airfield (Rugao) in central China, which turned out to be an exceptionally large aircraft depot to accommodate retired planes. Japan's Research for Peace and Security (RIPS), *Asian Security: 1998-1999*, p. 108, provides a figure of 3,740 warplanes, which is a reduction of 1,230 from 1997. The majority of the aircraft taken out of the inventory are the older J-6s, which were last produced in 1979. The actual figure is probably somewhere around 3,500 aircraft in the active inventory today.

36. There are no female fighter pilots in the PLAAF, but there are several female transport pilots.

37. Kenneth W. Allen, Glenn Krumel, and Jonathan D. Pollack, *China's Air Force Enters the 21st Century*, Santa Monica: RAND, 1995, p. 130.

38. Upon graduating from a basic flying academy, pilots move to a transition-training base, where training lasts for one year, 100 to 120 flying hours. The pilots begin flying the F-5 for basic airmanship, then transition to the F-6 or F-7. Upon graduation, the pilots are expected to be capable of flying in "three weather conditions", i.e., day and night visual flight rules (VFR), and day instrument flight rules (IFR). "Four weather conditions" adds night IFR flights. Thereafter annual flying hours vary according to the type of aircraft: bombers (80 hours), fighters (100–110 hours) and the A-5 ground attack aircraft (150 hours). *Dangdai Zhongguo Kongjun (China Today: Air Force)*, Beijing: China Social Sciences Press, 1989, pp. 503–504.

39. Cai Shanwu, "Build a Strong Air Force: An Interview with PLA Air Force Commander Wang Hai and Political Commissar Zhu Guang," *Renmin Ribao*, *FBIS*, November 10, 1989, p. 5. From "It is Hard to Give Full Play to Advanced Fighters," *Shijie Ribao*, August 22, 1997, A12, in John Wilson Lewis and Xue Litai,"China's Search for a Modern Air Force," *International Security*, Vol. 24, No. 1, Summer 1999, MIT, p. 86. This figure of 20.7 percent includes all officers, not just pilots. It is difficult to track the number or percentage of pilots in each category because the PLAAF does not routinely report these figures, or reports them in different forms, i.e., total pilots, not just fighter pilots.

40. Kenneth W. Allen, Glenn Krumel, and Jonathan D. Pollack, *China's Air Force Enters the 21st Century*, Santa Monica: RAND, 1995, p. 130.

41. Robert S. Dudney, "Fifteen in a Row," *Air Force Magazine*, April 1999, p. 25. Unfortunately, figures are not readily available for other Asian air forces.

42. Interview with a Department of Defense analyst.

43. Sun Maoqing, "Training Improves Air Force Combat Effectiveness," *Xinhua*, March 26, 1996. Sun Maoqing and Man Dongyan, "Air Force Flight Units Fulfill Training Mission," *Xinhua*, December 23, 1996. Neither of these articles attributed the excess training to the exercise conducted opposite Taiwan during March 1996 and the large exercise in the Gobi desert in September, but this is the most likely cause. Sun Maoqing, "Make Efforts To Build Modernized People's Air Force," Beijing *Liaowang*, No. 15, April 14, 1997, pp. 20-21. Zhang Nongke and Zhang Jinyu, "Air Force Builds Modern,

Comprehensive Tactical Training Base," *Jiefangjun Bao*, October 28, 1996. There are quite often conflicting stories that tell the amount of training that PLAAF flying units receive. For example, an article in Vol. 5 of the 1995 *Zhongguo Kongjun* (Air Force) magazine discusses the number of sorties flown (54,506) over the 8-year period of 1987 through 1994 by an unidentified air division on the Leizhou Peninsula. However, assuming the division has at least 72 aircraft, this equates to 6,813 sorties per year, or 94 sorties per aircraft per year, and less than two sorties per week.

44. Sun Maoqing, "Make Efforts To Build Modernized People's Air Force: Interview With Air Force Commander Lieutenant General Liu Shunyao," Beijing *Liaowang*, *FBIS*, No. 15, April 14, 1997, pp. 20-21.

45. "Deputy Commander Says Air Force Improves Combat Readiness," *Xinhua*, January 20, 1999, *FBIS-CHI*-99-020, January 20, 1999. These comments were made during a work conference on Air Force training.

46. Sun Maoqing, "Make Efforts To Build Modernized People's Air Force: Interview With Air Force Commander Lieutenant General Liu Shunyao," Beijing *Liaowang*, *FBIS*, April 14, 1997, No. 15, pp. 20-21.

47. This unit has been identified as the Blue Army (*Lan Jun*) and Blue Skies (*Lan Tian*) unit. The PLAAF visited Nellis Air Force Base in Nevada during the late 1980s and received briefings on the U.S. Air Force's Red Flag exercise.

48. In February 1987, the PLAAF Flight Test and Training Center (*kongjun feixing shiyan xunlian zhongxin*) was established in Cangxian (also known as Cangzhou), Hebei Province, replacing the 11th Aviation School.

49. Zhang Nongke and Zhang Jinyu, "'Grindstone' Confronts 'Iron Wings' in Blue Sky: Air Force Forms Unified 'Blue Army' Unit for First Time To Confront Airmen Units in Rotation," *Jiefangjun Bao*, April 28, 1997, *FBIS-CHI*-97-102, April 28, 1998.

50. Interviews.

51. Zhang Nongke, Niu Ruili, Liu Guojiang, and Guo Kai, "The Hypothetical Enemy of the Chinese Air Force," Hong Kong *Tzu Ching*, No. 106, *FBIS-CHI*-99-0824, August 5, 1999, pp. 19-23.

52. Interviews in Asia during May and June 1999 indicated that there were no Su-27s assigned to the aggressor squadrons. It is doubtful that any of the Su-27s assigned to the regiments at Wuhu or Suixi were

permanently assigned to the aggressor unit. The only logical explanation is that some of the first aircraft assembled at Shenyang have been assigned there. The first aircraft assembled at Shenyang flew in December 1998. An interview with a USAF F-15 pilot about this article indicated his concern about discrepancies in the article relating to the types of intercepts that were conducted. He cited a healthy dose of skepticism about the article's claims.

53. Historically, the PLAAF has not flown over water very much. This mission has been reserved for the Naval Air Force.

54. Department of Defense analyst.

55. Sun Maoqing, "Training Improves Air Force Combat Effectiveness," *Xinhua*, March 26, 1996, *FBIS-CHI*-99-018, January 18, 1999.

56. *Dangdai Zhongguo Kongjun (China Today: Air Force)*, Beijing: China Social Sciences Press, 1989, p. 311.

57. Zhang Nongke and Zhang Jinyu, "Air Force Builds Modern, Comprehensive Tactical Training Base," *Jiefangjun Bao*, October 28, 1996, *FBIS-CHI*-96-228, November 26, 1996.

58. *Ibid.*

59. "China Builds Airport Copy," *Associated Press*, April 28, 1999.

60. Zhang Nongke and Zhang Jinyu, "Summary of the Air Force's Pioneering Offensive-Defensive Confrontation Exercise With Participation of Varied Types of Aircraft (Arms of the Service) Under Unknown Conditions," *Jiefangjun Bao*, September 25, 1996, *FBIS-CHI*-96-205, September 25, 1996.

61. "PRC Air Force Upgrades Weather Forecast Equipment," *Xinhua*, January 13, 1998, *FBIS-CHI*-98-012, January 12, 1998.

62. Zhao Zhongwei and Fan Haisong, "Overall Combat Strength of the Air Force of the Guangzhou Military Region Crosses A New Threshold; First-Rate Equipment Builds An Aerial Great Wall of Steel," *Guangzhou Yangcheng Wanbao, FBIS*, July 22, 1999.

63. *Ibid.* This exercise appears to be an annual event for this bomber unit. According to *Zhongguo Kongjun* (China's Air Force), January 1989, Vol. 18, pp. 41–43, in September 1986 eight B-6 bombers from an air division in the Guangzhou MRAF participated in a 7-hour inter-MRAF long-range attack training exercise. The bombers first

dropped bombs at a bomb range in Hunan Province, then flew over three hours through five provinces to drop bombs at a bomb range in the northwest. Although the bombing results were good in Hunan, they were unsatisfactory in the northwest. The pilots were not familiar with the target, the sun was in their eyes, the target was difficult to find, and they did not arrive at the target on time, so they had to return home without dropping their bombs. As a result, they only received a rating of two for this portion of the exercise. In August 1987, another cell of eight bombers conducted a long-range attack based on a tactical scenario. The first target was hit during low level bombing, which was followed by a high altitude, long distance navigation route with a direct run on a target range on another lake. Before entering a simulated enemy radar net, they rapidly descended to quietly close on the target. Following this, they used maximum climbing speed to conduct their bombing. Before entering the bombing starting point, they met an unpredictable event. They were not able to see the number one target on an island in the middle of the lake because the water had risen. By the time the first two aircraft discovered this, it was too late to switch to another target. The third aircraft quickly switched to target number six on a peninsula on the lake. The bombs were dropped and hit 15 meters from the center of the target, resulting in a rating of five.

64. Zhao Zhongwei and Fan Haisong, "Overall Combat Strength of the Air Force of the Guangzhou Military Region Crosses A New Threshold; First-Rate Equipment Builds An Aerial Great Wall of Steel," *Guangzhou Yangcheng Wanbao* in Chinese, *FBIS-CHI*-99-0723, July 22, 1999.

65. "Storage Center for Fighters, Transport Planes, Bombers of Air Force," Hong Kong *Ta Kung Pao*, March 17, 1997. The U.S. Air Force hosted a small PLAAF delegation in the United States in 1988 to discuss maintenance and logistics. The delegation visited Hill Air Force base, Davis Monthan airbase, and the Smithsonian's aircraft refurbishment center in Washington, DC. One of the leaders of the delegation returned to Beijing and became the head of the Air Force's new aviation museum at Shahezhen airbase located just north of Beijing. During the visit, the delegation made known that they were planning to build a huge aircraft storage facility similar to Davis Monthan, but were having problems finding a place with the proper climate and necessary space. U.S. reconnaissance satellites discovered in June 1993 that China had gathered over 1,000 combat aircraft at Rugao airfield in central China. The storage center has three purposes: taking over and storing retired aircraft from all Air Force units nationwide; routine maintenance for those planes still functioning well; and renovating old and broken aircraft.

66. "The Security Situation in the Taiwan Strait," Report submitted by Secretary of Defense William Cohen to the U.S. Senate as directed by the FY99 Appropriations Bill, February 17, 1999. Although not cited here, there is an excellent paper by Michael Richard Danis entitled "China's Air Force in the Twenty-First Century: Chinese Acquisition and Production of Advanced Aerospace Technology," that appeared in *The China Strategic Institute*, Vol. III, No. 1, Spring 1998, pp. 122-135.

67. According to a Department of Defense analyst, the J-8B (F-8II) and J-8D have a forward hemisphere BVR engagement capability with the PL-4/PL-11 (*Aspide/Aspide* copy) AAM. The J-8D has a forward hemisphere self-protection jammer. The J-8IIM development program with Russia will be completed in 2000. It includes the ZhUK-8II pulse doppler radar and AA-10A and AA-X-12 active AAM. The J-8IIM has better BVR capability than the Su-27SK in the Chinese inventory. The upgrade also includes the ability to fire KH-31A antiship missiles. The Chinese could produce the J-8IIM or upgrade the J-8B/D, or both. The Chinese currently have approximately 100 J-8B/D deployed. While the J-8IIM is an export program, the PLAAF is weighing its options on this program.

68. According a Department of Defense engineer with over 10 years experience analyzing Chinese aircraft, the J-7 and J-8 have several variants with modern subsystems. For example, Chinese promotional brochures describe the J-7C/D as having a monopulse fire control radar capable of off-boresight engagements and can accept data link inputs. The J-7E, with its larger cranked wing with leading edge flaps is capable of withstanding 8 Gs, and is therefore quite maneuverable and lethal. The J-7D, J-7E, and J-8D all have an upgraded radar warning receiver that would allow them to better detect Western aircraft. Combined with improved electronic countermeasures (ECM), they could close to within visual range (WVR) for air-to-air combat. The Chinese are working several simple upgrade programs that could further improve their near-term capabilities. The J-7MG is nearing completion, and has a pulse doppler radar called the Super SkyRanger. It is still WVR, but it has ACM modes and look-down capability. At the Zhuhai airshow, the Chinese claimed that all aircraft would be upgraded with a helmet mounted sight (HMS). The Russians have proven the value of HMS. They cut typical lock-on time from 5-15 seconds, to 1-2 seconds, even at high off-boresight angles. They also claimed that the KG-300 jammer pod was tested on a J-7. If you upgrade J-7MG with the *Super SkyRanger*, and HMS, and the KG-300 jammer pod, J-7E would be a very capable aircraft.

69. Kenneth Allen, "PLAAF Modernization: An Assessment," in *Crisis in the Taiwan Strait,*, Washington, DC: National Defense University Press, September 1997, pp. 218-245.

70. Hsiao Yu-sheng, "China's New Generation Main Military Aircraft," *Kuang Chiao Ching*, Hong Kong, November 16, 1995.

71. "First Flight for F-10 Paves Way for Production," *Jane's Defence Weekly*, May 27, 1998, p. 17.

72. This agreement was important psychologically to China and to China's neighbors. Not only was China going to receive a modern aircraft with offensive capabilities, but also it meant that Russia might be willing to sell China other modern weapons.

73. "China-assembled Su-27s Make Their First Flights," *Jane's Defence Weekly*, Vol. 31, No. 8, February 24, 1999.

74. Based on interviews in Asia, the PLAAF has configured a handful of B-6 bombers as tankers and is currently using them to refuel J-8II interceptors in the Guangzhou Military Region.

75. According to several reports, the current contract stipulates that the Chinese share of components will only reach 70 percent. The AL-31F turbofan engine and various avionics are not among the items that will be license produced by China. The contract also prohibits any changes ever to the Su-27s or any exports without Russian authorization. Nikolay Novichkov, "China To Begin Licensed Production of Su-27CK Fighters," *Itar-Tass*, in English, November 13, 1997. "Licensed Building of Russian Fighters about to Begin in China," *Itar-Tass*, October 2, 1998. Although the first report identifies the Su-27 variant for China as the Su-27CK, the second report identifies it as the Su-27SK. While some sources who have toured the Shenyang factory state that manufacturing capabilities and procedures are generally poor and could take several years before successfully assimilating Russian manufacturing technologies, other sources have stated that the Su-27 is not an advanced aircraft from a manufacturing standpoint. It is an all metal airframe with only limited exotic materials/manufacturing processes, such as titanium components, which might cause some difficulties. Each is hand built and is not significantly more advanced than the J-8 airframe built at Shenyang. A clause in the SU-27 contract stipulates that if Shenyang fails to meet the annual production target of 10 to 15 aircraft, then Russia's facility at Komsomolsk will provide the substitute aircraft. Even though China is not allowed to produce the AL-31F engine, which is also being used for the J-10, China is trying to purchase an engine repair maintenance facility, so that they do not have

to continue sending the engines back to Russia for repair, "Licensed Building of Russian Fighters about to Begin in China," *Itar-Tass*, October 2, 1998. According to some analysts, the Chinese have not asked for licensed production yet of the AL-31F engine and existing avionics incorporated in the Su-27 because they are waiting to see if their indigenous engine and avionics programs will be successful. China's domestic development programs in these areas are based on purchases of or negotiations for some engine and avionics technology that is more advanced than what is in the Su-27, according to a Department of Defense analyst.

76. "'Made in China' deal is forged for Su-27s," *Jane's Defence Weekly*, Vol. 23, No. 18, May 6, 1995, p. 3.

77. Department of Defense engineer.

78. According to an August 19, 1998, *Jane's Defence Weekly* article, China and Russia signed a memorandum of understanding to open formal negotiations on China's proposed purchase of more than 20 SU-30s from Russia and is discussing the transfer of production technology. The latest report from the *Central News Agency* out of Hong Kong on June 21, 1999, and *Defense News* says that Moscow has decided to sell 50 to 72 of its front-line SU-30 jet fighter-bombers to Beijing. The cost of the aircraft will be $37 million a copy. In addition, the *Central News Agency* reported that negotiations for Moscow to grant a license for the production of another 250 SU-30 fighters in China have begun as well. Like the wide range of reporting over the past 9 years on the Su-27s, it will still take some time before the actual terms of the contract are available.

79. Department of Defense engineer. The Su-30MKK does not include thrust vectoring or the Su-30s distinctive canard. Information from the air show suggests that the Su-30MKK has an upgraded Slot Back2 to allow air-to-ground operations and air-to-ground munitions. The only other big difference over the Su-27SK is that it does include aerial refueling. Therefore, this aircraft will be able to perform deep strike missions normally performed by B-5 *Beagle* bombers, yet still retain air superiority capabilities, making it similar to the F-15E.

80. "The Security Situation in the Taiwan Strait," Report submitted by Secretary of Defense William Cohen to the U.S. Senate as directed by the FY99 Appropriations Bill, February 17, 1999.

81. In 1956, the Soviets and Chinese signed a contract to build a medium bomber factory at Xian, and in September 1957, the Soviets gave China the rights to produce the TU-16. In February 1959, the

Soviets turned over the plans for the TU-16, as well as two aircraft and one unassembled aircraft to Harbin, where the first flight took place in September 1959. The Xian production facility was completed in 1961, and the TU-16 production capability was transferred there from Harbin between 1962-1964. Strength tests of the first indigenously produced B-6 were completed in October 1966, and the first test flight took place on December 24, 1968. The B-6 entered series production in 1969. Development of the B-6D began in 1975, and the first flight took place on August 29, 1981. On December 6, 1981, it conducted telemeter bomb tests at a bomb range, and conducted live testing at the end of 1983.

82. "The Security Situation in the Taiwan Strait," Report submitted by Secretary of Defense William Cohen to the U.S. Senate as directed by the FY99 Appropriations Bill, February 17, 1999.

83. *Ibid.*

84. Interview with Ministry of Defense officials in New Delhi.

85. "Delivery of China's upgraded Il-76 AEW system is delayed," *Flight International*, August 19-25, 1998, p. 21.

86. According to one Department of Defense analyst, the probability of success for the Il-76/Phalcon program will be quite good, especially if the Israelis provide support as in the J-10 program. The Phalcon system makes extensive use of commercial-off-the-shelf (COTS) products, which gives easy access to the basic building blocks of the system. The system is extremely capable and has DOD officials concerned. The Phalcon system was a competitor in the recent Australian Wedgetail AEW competition. The system is much more than an AEW and has full AWACS capabilities. In addition, the Wedgetail proposal included Australian industrial participation. Therefore, it is not unreasonable to believe that the Israelis offered the Chinese industrial participation to seal this high dollar deal.

87. Desmond Ball, "Signals Intelligence in China," *Jane's Intelligence Review*, Vol. 7, No. 8, August 1, 1995, p. 365.

88. "Elint aircraft put into service," *Jane's Defence Weekly*, Vol. 31, No. 12, March 24, 1999. Based on discussions with various analysts, there are indications that the information about the Tu-154s deploying to the Nanjing Military Region may be an attempt at disinformation.

89. "The Security Situation in the Taiwan Strait," Report submitted by Secretary of Defense William Cohen to the U.S. Senate as directed by the FY99 Appropriations Bill, February 17, 1999.

90. "NAF regt improves new aircraft," *People's Navy*, June 18, 1998.

91. Department of Defense analyst.

92. "The Security Situation in the Taiwan Strait," Report submitted by Secretary of Defense William Cohen to the U.S. Senate as directed by the FY99 Appropriations Bill, February 17, 1999.

93. Min Zengfu and Han Jibing, "Air Defense Pattern of the 21st Century: Large-Area Dynamic Joint Air Defense," *Jiefangjun Bao*, July 14, 1998.

94. *Ibid*.

95. Sun Maoqing, "Make Efforts To Build Modernized People's Air Force: Interview With Air Force Commander Lieutenant General Liu Shunyao," Beijing *Liaowang*, *FBIS*, April 14, 1997, No. 15, pp. 20-21.

96. "China seeks S-300 and Tor-M1 systems," *Jane's Defense Weekly*, September 2, 1998, p. 16. The PLA's S-300MU1 can intercept aircraft-sized targets at ranges up to 150 kilometers and ballistic type targets at ranges up to 40 kilometers. The Tor-M1 is a low-to-medium altitude self-propelled SAM which entered Russian service in 1991.

97. "Chinese Exercise Strait 961: 8-25 March 1996," briefing presented by the U.S. Office of Naval Intelligence at a conference on the PRC's military modernization sponsored by the Alexis de Tocqueville Institute, March 11, 1997.

98. "China shows new AWACS-killer missile," *Jane's Defence Weekly*, Vol. 30, No. 11, September 16, 1998. Paul Beaver, "China flexes muscles in missile exercise," *Jane's Defence Weekly*, Vol. 32, No. 5, August 4, 1999.

99. "PLAAF aircraft target Taiwan Aircraft," *United Daily News*, August 5, 1999. "Taiwan Air Force Monitoring China Air Force Movements," Taipei *Chung-Yang Jih-Pao* in Chinese, *FBIS*, August 4, 1999, p. 10. These reports stated: (1) Two PLAAF J-8 fighters crossed the centerline of the Strait on July 25 and two J-7 fighters crossed the centerline on July 30. Shortly after the planes crossed the line, they corrected their routes and returned to the west side of the center line; (2) The J-7s and J-8s from Fuzhou and Longxi airbases were flying over the Strait on a daily basis; (3) Su-27 fighters stationed at Suixi airbase in Guangdong Province began training activities in areas close to the center line. (The Su-27s from Wuhu airbase in Anhui Province were probably also involved); (4) There was no truth to reports that PLAAF Su-27s locked on with their fire control radar to two Taiwan Mirage

2000 fighters over the Taiwan Strait on August 2; and 5) About 340 international flights and 730 domestic flights fly over the Taiwan Strait daily.

100. Victor Lai, "PRC Jet Fighters Twice Cross Taiwan Strait Center Line," *Central News Agency,* Taipei, August 10, 1999.

101. Lo Ping, "It Costs China 3 Billion Yuan to Make a Show of Its Military Strength," *Cheng Ming*, Hong Kong, April 15, 1996.

102. "Chinese Exercise Strait 961: March 8-25, 1996," briefing presented by the U.S. Office of Naval Intelligence at a conference on the PRC's military modernization sponsored by the Alexis de Tocqueville Institute, March 11, 1997.

103. Ping, "It Costs China 3 Billion Yuan to Make a Show of Its Military Strength"; Steven Mufson, "China Masses Troops On Coast Near Taiwan," *The Washington Post*, February 14, 1996.

104. "Mainland Fighters Said To Appear Above Taiwan Straits," Taipei *Tzu-Li Wan-Pao,* November 26, 1998, *FBIS-CHI*-98-333, November 29, 1998.

105. State Department Briefing with James Rubin, *Federal Information Systems Corporation, Federal News Service*, August 3, 1999.

106. Interview with government official.

107. Lu Xingsheng, "Taiwan Haixia Cong Wei Cun Zaiguo Shenma 'Zhong Xian'" (The Taiwan Strait has Never Had a Center Line), Shanghai *Xinmin Weekly*, No. 34, August 23-29, 1999, p. 8.

108. Taiwan's Air Force has now received all of its 130 IDF, 60 *Mirage*, and 150 F-16 fighters.

109. State Department Briefing with James Rubin, *Federal Information Systems Corporation, Federal News Service*, August 3, 1999.

110. *Republic of China: 1998 National Defense Report*, Taipei: Li Ming Cultural Enterprise Co., Ltd. 1998, pp. 184-185. Almost none of these aircraft have any standoff weapons launch capability. Therefore, they would not be that effective in reaching Taiwan unless China were to take out most of Taiwan's air defense capability with ballistic and cruise missiles first.

111. "Chinese Exercise Strait 961: March 8-25, 1996," briefing presented by the U.S. Office of Naval Intelligence at a conference on the PRC's military modernization, sponsored by the Alexis de Tocqueville Institute, March 11, 1997.

112. Sergey Sokut and Ilya Kedrov, "War in Europe: Yugoslavia: 78 Days Under Missile and Bombing Attacks: NATO's Limited War Was Uncompromising," Moscow *Nezavisimoye Voyennoye Obozreniye, FBIS*, No. 25, July 2-8, 1999, p. 2.

113. Kenneth W. Allen, *People's Republic of China, People's Liberation Army Air Force*, Washington, DC: Defense Intelligence Agency, May 1991, Section 2.

114. "Taiwan Air Force Monitoring China Air Force Movements," *Taipei Chung-Yang Jih-Pao*, Internet, August 4, 1999.

115. Kenneth W. Allen, Glenn Krumel, and Jonathan D. Pollack, *China's Air Force Enters the 21st Century*, Santa Monica: RAND, 1995, p. 64.

CHAPTER 7

THE KOSOVO WAR: IMPLICATIONS FOR TAIWAN

Arthur Waldron

Kosovo will not be remembered as one of NATO's signal successes. That is bad news, for the long-standing People's Republic of China (PRC) plans to use force in a similar manner to bring Taiwan under Beijing's control. Nearly a year after North Atlantic Treaty Organization (NATO) sorties ceased, the Serbian government of Slobodan Milosevic remains in power, and the situation on the ground in Kosovo is as volatile as ever. More troops may be sent there, but no solution is in sight.

The lessons of NATO's Kosovo campaign, both diplomatic and military, throw into doubt the fundamental strategy that Beijing has long employed against Taiwan. The strategy is pure realpolitik of the type familiar during most of the 20th century. Diplomatically, the goal was to isolate and delegitimize the island globally by securing recognition of China's absolute sovereignty from the leading powers, and to pressure the government in Taipei with the assistance of the United States, in particular. Held in reserve was a military threat, including nuclear weapons, not only against the people of the island, but also against any country that might have considered aiding it. Until recently, this strategy worked quite well.

But, as Beijing realizes, the Kosovo campaign throws these calculations into question. From the diplomatic point of view, Kosovo suggests that the concept of absolute sovereignty is weakening. No one doubted, after all, that Kosovo was part of Serbia. Increasingly, it seems, as a Canadian diplomat put it, "Sovereignty is no excuse for human rights abuses."[1] As for the military, the prospect that a few well-placed missiles might bring Taipei to heel faded dramatically when the flurry of Tomahawk missiles U.S. Secretary of State Madeleine Albright hoped would nudge Milosevic to talks had no such effect. Instead, there ensued an inconclusive 78-day bombing campaign, in which the full might of NATO, stronger by any measure than China, was hard put to dominate little Serbia and its dictators' dictator—a smaller and far weaker target, by any measure, than democratic Taiwan.

Not surprisingly, the rhetoric from Beijing became almost frantic as the story unfolded. The attack published in the *People's Daily* of June 22, 1999, perhaps takes the prize—its (unfavorable) comparison of the United States to Hitler is as scurrilous as anything to come out of Beijing since the Korean War:

> ... Hitler not only used in war what were considered to be the most advanced weapons of the time, such as airplanes, tanks, and long-range artillery, to massacre peaceful citizens in anti-fascist countries, but also built concentration camps in Auschwitz and in other areas to slaughter Jews and prisoners of war with "advanced" technology. Executioners drove hundreds and thousands of people into gas chambers and poured cyanide through air holes in the roof, killing them all. Today, the U.S. hegemonists use high-tech weapons to attack the FRY's [Federal Republic of Yugoslavia] civilian facilities, several hundred miles away from the battlefield or with laser and global positioning system several thousand meters in the sky, treating innocent and peaceful citizens as live targets. The flagrant use of missiles by the U.S.-led NATO to attack the Chinese embassy in Yugoslavia was a barbaric atrocity that the then Nazi Germany had not dared to commit.[2]

But Beijing's response has been confused: praising Milosevic (but not really helping him, even in the U.N.), attempting a speedy reconciliation with Russia (which, like China, has no desire to alienate the West), adding thick layers of theatricality and menace to quite justified indignation at the mistaken U.S. bombing of the Chinese embassy, playing hard to get over World Trade Organization (WTO) accession, and other issues.

China senses that Kosovo is an indication of how thoroughly the post-Cold War world is failing to fulfill her hopes. Given that China draws strength from effective operations in the international environment, this poses a serious problem—and one to which there is no solution in the current framework of Chinese strategic thinking.

Diplomacy.

China is deeply concerned by the way the international law of sovereignty, on which Chinese governments throughout this century have staked so much, is beginning to incorporate troublesome notions such as self-determination and human rights.

The first line of defense is to deny that human rights abuses exist. "Remarks and reports on Yugoslavia's so-called 'ethnic cleansing' have different interpretations," said Zhu Bangzao, foreign ministry spokesman.[3] The West has been spouting "misleading and demagogic propaganda about the FRY's 'ethnic cleansing' and 'oppression of human rights' in Kosovo." The International Court of Justice has even declared Milosevic a war criminal and "openly appealed to the FRY people to overthrow the FRY government"—and, as final twist of the knife, threatened to withhold financial aid from Serbia until Milosevic is replaced.[4] But acknowledgment of the truth had appeared as early as April 20, 1999, in the *China Economic Times*: "on the road to exile, the Serbs believe everything is permitted. They are raping women, killing and robbing refugees, no one can stop them"—it quoted a refugee as saying.[5]

China today engages in appalling repression against its own people, minorities included, and threatens the same to Taiwan, all under the rubric of sovereignty. But if Kosovo is a precedent, the world community could conceivably condemn Beijing for this. They might even indict some

Chinese officials, or call upon the Chinese people to overthrow their designated leaders—or (perhaps the most realistic worry)—refuse money (as they are refusing it for the Three Gorges Dam) out of some cockamamie concept of freedom.

The prospect is particularly worrying because, throughout this century, China, when weak, has drawn strength from international law, which it has attempted to use to unify itself. Thus in the 1920s, the provisions of the Washington treaties were distorted into a foreign guarantee of Chinese unity. In the 1930s, international law was used powerfully against Japan, and since the 1970s, China has used the U.N. and its legal structures to stifle and isolate Taiwan, just as Taiwan used them against China until it left the organization in 1971. China in the past has used international law to leverage its own claims. Now that legal system is starting to be used to leverage other sorts of claims.

Most important here are "so-called 'democracy' and 'human rights'" and the advocacy of "fighting for values [instead of territory]" to which "the United States and its allies now give wide publicity." This is directed against socialist countries as well as developing countries that are not socialist but that do not obey the United States. Moreover, the United States uses "common values" as a bonding agent for consolidating its relations with its allies.[6]

We have an odd blend here. Note the reaffirmation of socialism, jarring in a China that is supposed to be reforming. And the sense that the socialist camp is beleaguered. But there is also an odd sort of caricature "realism" here, odd given the ideological definitions also used. Countries are supposed to fight over territory and resources, not "values" (though socialism is presumably a value). And there is a recognition that, despite all the rhetoric about U.S. military power being the source of its hegemony, actually "common values" have something to do with it.

The threat, moreover, is spreading. The U.N. used to be a reliable bastion for third world dictators and communists, but now a lot of democrats have turned up there, so much so that the U.N. itself ended up endorsing the Kosovo operation. And the "heads of NATO countries have even claimed that if the United Nations does not allow itself to be manipulated, they will separately form a so-called 'Alliance of Democratic Countries'"—in which China presumably would not even have a seat, let alone a veto.[7]

The U.N. used to insist on absolute sovereignty and noninterference, as a series of declarations in 1949 and 1965 made clear. Now it is abandoning that. On June 10, 1999, the Security Council passed a resolution legitimizing the bombing campaign and the occupation of Kosovo—and China only abstained. Rhetoric, obviously, and pretty poor at that—but not without a kernel of truth, which is that the post-Cold War world has developed in a way entirely different from what China expected, and that is most unwelcome. China expected NATO to dissolve, to be replaced by a series of power centers—Europe, Russia, Japan, and so forth—among which China could work its usual diplomatic stratagems of ever shifting orientations. But instead, NATO not only survived, it expanded—and it found as its new mission something to do with human rights. Its ambitions became global. Conquest of the "strategically important Balkans" (strategically important for what?) would permit NATO to move east, "eventually gaining control of the whole Europe, including Russia." That Western pincer would be matched by the Alliance with its "important accomplice, Japan" and culminate in dominating the world.[8]

This led to the position taken by China early in the conflict that Serbia was the "red fortress of Europe," waving the red flag heroically against the "so-called 'Western democratic community';" Milosevic was a doughty worker's hero, a "second Tito," and "A man of iron who does not bend to the West, and a folk hero who can make his nation follow

259

him with a single call or raise of an arm" (better than Jiang Zemin?)[9]—characterizations that were sadly disappointing.

But in practical terms, China did very little to help the Serbs. Pro-Serbian rallies in Beijing were halted, and when it came to the crunch in the U.N., the dreaded Chinese veto shrank into an abstention. "The Chinese are ready to defend Serbia down to the last Russian," one diplomat observed.[10]

Implications for Taiwan.

In Switzerland, Jiang Zemin referred to NATO's actions as "new gunboat policy," and the Chinese Communist Party (CCP) issued internal circulars on the "hegemonic acts" of the United States in the Balkans. Revising the so-called pro-U.S. foreign policy initiated by Deng Xiaoping, senior Politburo members drew comparisons between Kosovo on one hand, and Taiwan and Tibet on the other, in closed-door sessions.

> If U.S.-led foreign powers could wantonly interfere in Yugoslavia in support of a pro-independence movement, the precedent could conceivably be used one day with reference to Taiwan, Tibet or Xinjiang.[11]

The intervention demonstrated that in the world today, moral and humanitarian considerations favoring self-determination and opposing terror and ethnic cleansing clearly trump old-fashioned absolute sovereignty of the sort that China claims over, say, Tibet (where it has certainly committed appalling atrocities) or Taiwan.

This suggests that Taiwan's claims to democratic legitimacy will have an impact even on a world that has largely acquiesced in China's claims of sovereignty, to the extent that China could scarcely expect support from anyone if it were actually to attack Taiwan, while many states would cooperate with the few who might actively support the island.

In other words, even given sovereignty, a military solution would be sufficiently unwelcome to the rest of the world that it might not work at all or, even if it did, be accepted. We see signs of this sort of attitude in the international reaction to Taipei's "liang guo lun" [two states theory] of Republic of China (ROC)-PRC relations—i.e., like the former two Germanys, they are two states within one nation. Condemnation has come from China, predictably enough, from the U.S. State Department, and from some American China watchers. But the fact that Taiwan is a democracy enters into it increasingly, as in a letter to the *New York Times*:

> . . . you correctly point out that the Taiwanese exercise democracy; on the other hand you fail to consider the possibility that President Lee Teng-hui's proclamation of a "state-to-state" relationship to China reflects longtime public opinion in Taiwan.[12]

This can only be worrying to China. In the past, Beijing has managed to avoid condemnation on the grounds of human rights by blurring the issues, baiting foreigners with the prospect of financial gain, and issuing threats. It was on the latter that China seemed to move most actively after the Kosovo war. But here again, the lessons were not encouraging.

The Use of Force.

Force would seem to have become a more attractive option for China's leaders in the years since the Tiananmen Massacre and Deng Xiaoping's incapacitation and death. Deng was strong enough at home to recognize China's weakness abroad and to understand that her true long-term interests would be served by peace and economic development. This did not stop him from promoting nuclear and missile programs, or even from invading sovereign Vietnam, but it did give him immunity to the appeal of xenophobia, nationalism, and the "splendid little war" as tools of domestic statecraft.

His successors are quite different. They are weak and divided at home. They could seek legitimacy through reform but understand that would mean abolishing their own jobs. So, not surprisingly perhaps, they have turned to nationalistic slogans at home and saber rattling abroad.

One lesson of Kosovo (and of the collapse of the Union of Soviet Socialist Republics) is that, according to Wang Yizhou of the World Economics and Politics Research Institute, "We cannot be soft-hearted or tolerant toward any separatist groups, including those in Tibet, Taiwan, and Xinjiang."[13]

Undergirding this has been an assessment of the international community and of the utility of force. Until recently, the Chinese view has been that the post-Cold War international community was resolving itself into a group of competing blocs, among which China could play its games of triangulation. Until recently as well, the Chinese view has been that ballistic missiles and nuclear weapons could form the foundation of a military force that would almost guarantee a form of hegemony in Asia.

These two big facts would permit the regime to notch up some important achievements in the years ahead. Taiwan would be intimidated into settling or be conquered; Japan would be nobbled militarily and forced to acknowledge Chinese primacy; and Korea would become a close friend, the South China Sea a Chinese lake, and the United States a presence offstage somewhere around Hawaii. Kosovo turned all these calculations upside down.

The PRC government faces a plate full of problems, of which Taiwan is rather minor. (Indeed, given all the money Taiwan pours into China and the market it provides, an independent Taiwan is actually a net plus). A "splendid little war" with Taiwan might look like just the ticket to soothe some of these problems, by diverting attention away from the undemocratic system, the corrupt leadership, and so forth, toward foreign issues. But as will be seen, such a war would be neither splendid nor little.

In this respect, if China launched such an attack on Taiwan, it would be in a similar position to NATO, where most of its people cared not at all about what was going on in Kosovo. That fact meant that NATO had to fight a very safe war and ask for no sacrifices. Its leadership was moreover undermined by President Clinton's well-known evasion of military service. Had Milosevic been able to follow-up on his successful downing of an F-117 plane or inflict other dramatic damage on NATO, the Alliance would most likely have called off the war.

The same sort of dynamic would apply in China. The levels of pilot and other losses one can reasonably anticipate, not to mention some strong counterstrikes from Taiwan, would probably convince most Chinese that the island was not worth the bones of a single "Shandong Hao Han." Furthermore, as happened in NATO, as soon as a Taiwan crisis protracted, political authority would be stretched to the limit, all the more so because in China there are no legitimating mechanisms for authority. A quick victory would win laurels for the central government, but setbacks would translate into defections of all sorts: local leaders would tilt away and withhold funds; political rivals would make hay; and insurgents in Xinjiang might try something. The whole Chinese dominance hierarchy could begin to collapse into anarchy. (War would have the opposite effect in Taiwan, where public opinion would be solidified).

Remember there is an asymmetry in interests. China has NO vital interest in Taiwan: the island poses no threat whatsoever to her, unless we consider the example of democracy to be a threat to a dictatorial regime, and that is not a military threat. Far more serious challenges to China's military position are posed currently by India's accelerating development of nuclear forces and a blue water navy, and South Korea's growing submarine force and increasing numbers of missiles, and in the future by a possible Russian turn against China and Japan's shift to military great power status (which looks likely to be accelerated by current

Chinese moves). To be sure, the current Beijing leadership places great emphasis on Taiwan as *terra irredenta*, part of the motherland not yet redeemed (though it makes no such claims on Mongolia (whose independence China recognizes), even though Mongolia is arguably at least as important strategically and was part of the Qing Chinese empire far longer than was Taiwan. Without dismissing the national unification concerns entirely, it can perhaps be noted that neither Mao nor Deng, both nationalists, were anywhere near so insistent as is the current leadership. Indeed, one may suspect that the rather weak Jiang Zemin leadership sees truculence over Taiwan as a way to rally nationalistic opinion and—who knows?—possibly even persuade the United States to lean even harder on Taipei to come to terms. And surely they understand that a war over Taiwan would be popular only if it was easy and succeeded quickly. To expend substantial blood and treasure there would be unpopular. Abandonment of such a venture, by contrast, would cost little and be accepted easily. Taiwan, however, would have no less than a SURVIVAL interest: defeat would mean enslavement and loss of independent existence. What a Finnish student once remarked at the Naval War College also applies to Taiwan: "Our country does not have the concept of limited war."

In this respect, the issue of force against Taiwan is for Jiang Zemin not unlike what the issue of force against Kosovo was for Clinton. Arguably, Clinton's lack of military experience and consequent need to prove himself made him more likely to use force against Kosovo, particularly after NATO painted itself into a corner over the negotiations. Neither Mao nor Deng had any such need to prove themselves, but Jiang, who lacks any military experience but relies on the military as the ultimate prop of his rule, may feel one. Like Clinton, however, he will be attracted only to what looks to be an easy, short, and relatively bloodless conflict. Some in the Chinese military speak of winning a war over Taiwan in 24 hours (the plan for Kosovo was 48). Hence all the more reason for Beijing, above all, to take note of what really happened: how badly the war went compared to expectations, and how much damage it did to the credibility of both NATO leaders and forces.

The Allied Operation.

Although it would be stretching things to call the Kosovo operation a success, what the Chinese saw operationally was undoubtedly troubling. By Chinese standards the Serbian military is by no means obsolete, yet it proved powerless to interfere with the allied air campaign, not to mention incapable of striking back against their attackers in any way. True, things might have been different had NATO moved in on the ground, attacking entrenched positions, but that proved unnecessary.

The U.S. Army would probably have proved too "heavy" to be brought into the operation, had that been necessary. But the new weapons that will form the core of the U.S. military in a decade's time, moreover, did very well. The problem was not enough of them. Boeing's Joint Direct Attack Munitions (JDAM), Raytheon's AGM-154 Joint Standoff Weapon (JSOW), and the Lockheed Martin Joint Air-to-Surface Standoff Missile (JASSM)—all so-called "smart weapons"—did well enough that Pentagon planners intend to give them a primary role in U.S. forces. The B-2A Spirit also performed unexpectedly well.[14]

One problem was the NATO shortage of so-called "low quantity—high utilization" weapons, such as the Airborne Warning and Control System (AWACS), Joint Surveillance Target Attack Radar System (JSTARS) and U.S. Navy and U.S. Marine Corps EA-6B Prowler jamming aircraft. Compared to China, NATO's order of battle contains large numbers of these elements, but scarcely enough to deal with the challenges posed by little Serbia even with units pulled in from all over the globe, including Asia. Even so, rather rudimentary Serb deception proved highly effective. Alliance pilots may have destroyed far fewer tanks, for example, than previously thought, because so many of the targets were decoys and were also difficult to hit.[15]

What would all this mean for a potential Taiwan operation? The island would be an incomparably tougher target than was Serbia. NATO has far greater electronic warfare capabilities than China, and the Republic of China (Taiwan) countermeasures are far better than what Serbia possesses. Yet NATO was unable to completely suppress its adversary. That bodes ill for any Chinese operation.

Furthermore, Serbia had no ability to strike NATO bases and anti-aircraft capabilities that are far inferior to the ROC. Yet the Serbian air threat environment was sufficient to force NATO aircraft to fly too high for optimal operations. What NATO was unable to do against Serbia in 78 days, China is unlikely to be able to do against Taiwan.

Second, the style of military intervention that China increasingly envisions proved highly ineffective in the Kosovo war, even when carried out by NATO's experts and against a state far smaller and weaker and isolated than is Taiwan.

Secretary of State Albright, it will be recalled, sought in the Kosovo case to nudge Milosevic back to the negotiating table at Rambouillet by means of a whiff of Tomahawk missiles. These were of course duly fired, but, instead of scurrying back to France, Milosevic sat tight, and the NATO campaign grew steadily for 77 days, until more than 600 aircraft were involved. A great deal of damage was done—but still, Milosevic yielded only when he was offered far better terms than the Rambouillet deal.

This experience merely confirms again what we have learned repeatedly from World War II to Vietnam to the Gulf War—bombing alone is not enough to win, let alone win quickly. This is particularly true when conventional warheads are used: they are relatively weak compared to the cost of the missiles. Hitler's V-2s were a technological marvel, but they no effect on Britain.

If we apply the Balkan parallel almost exactly to the Taiwan Strait, the scenario would be something like this. China and Taiwan reach an impasse in negotiations, and Taiwan goes home. PRC warns and then begins firing missiles at the island. After 78 days, PRC then backs down and compromises with Taiwan.

That would be bad enough given the Tiananmen Square standard for obedience that still prevails in China. But even that outcome is unlikely. For as Clausewitz reminds us, in warfare, reaction is correlated to action—not in fixed mathematical fashion as in physics, but rather "actively," which is to say, unpredictably.

China could not deliver the kind of sustained punishment to Taiwan that NATO did to Yugoslavia and, furthermore, even if no other states intervened, Taiwan would be unlikely simply to accept 77 days of pounding as Yugoslavia did. It would react by striking back at China with great destructiveness.

Defense.

NATO was unable to effectively cripple even Yugoslavia's relatively simple air defenses; for a much weaker China to destroy the air defenses of a much stronger Taiwan would be far more difficult. Certainly large numbers of pilots would be lost, and, even then, victory would be most uncertain. The same is true for maritime operations.

The Revolution in Military Affairs (RMA) only intensifies this reality. A first-hand observer reports that damage by NATO's precision guided munitions to Yugoslav targets was comparable in many cases to what could only have been achieved by nuclear weapons in the past—yet almost without collateral damage. Key nodes of every sort were effectively disabled at relatively low cost.

Waging an offensive like NATO's requires large and vulnerable bases, logistical trains, communications, and so forth. Had Yugoslavia possessed some cruise missile and other RMA sorts of capabilities, it could have wreaked havoc on NATO bases in Italy and elsewhere.

Massing troops and aircraft for an operation against Taiwan would be a protracted and difficult business that could be crippled by precision strikes against transportation and other infrastructure. What was intended to be a lightning strike could bog down and protract. A worst case for the PRC would be a repeat of the Guningtou battle that destroyed a Chinese expeditionary force attacking Jinmen in October 1949.[16] The difference being that, whereas 50 years ago the PRC forces were trapped on the beaches of Jinmen and annihilated by ground fire after their landing craft had all been sunk, in this scenario the expeditionary forces trapped in Fujian Province would be cut off from the rest of China and then destroyed from the air.

Theater Missile Defense.

China wants great-power-style influence, but it cannot afford great-power-style forces, nor is it willing to adopt the cooperative diplomatic norms that give the United States or Great Britain their real influence. So it has fixed on ballistic missiles and nuclear weapons as shortcuts to such influence, massively investing in such systems. The problem is that, even in the best of times, the ballistic missile with nuclear warhead is useful chiefly as a deterrent, to keep someone else from attacking you, rather than as an instrument of coercion—given that the threat to drop atomic bombs on someone with whom you simply disagree seems far too risky to be at all plausible.

Those limitations are now compounded by two others. First is that of countervailing moves and retaliation. The sort of missile and precision guided munitions threat being discussed here is not Taiwan-specific. It potentially threatens any state within China's range, and such states will behave accordingly. It is hard to imagine that Japan, Korea, Taiwan, and other states will simply stand by as

China deploys its DF-31s. They will certainly speed up development of their own missile forces.

Add to this is the dawning U.S. recognition that missile defense is essential, and progress is being made in, for example, the Navy theater-wide system. This system could be deployed on ships to threatened areas and destroy a respectable percentage of missiles in flight. That fact would greatly reduce both the psychological and the military impact of a missile barrage. Not surprisingly, PRC opposes it. But the lesson of Kosovo is that you need such a system, otherwise you are naked to your adversary, and this lesson is not lost on the other Asian states. China's renewed threats early in 2000 to use missiles only remind the region of the threat, and make defensive action even more likely.

Collateral Aspects.

Even in the Yugoslav case where civilian deaths were relatively limited, NATO public opinion was deeply troubled by innocent deaths. In a Taiwan Strait scenario, collateral damage would be far greater and also better documented. PRC would be attacking a Taiwan full of local and foreign journalists, as well as diplomatic representatives from all over the world, most of them deeply sympathetic (as was not the case in Belgrade) with the plight of the locals, whatever their governments might think. Large amounts of foreign property and investment would also be at stake. A great deal would be covered in real time by CNN and other such outlets. Not only would this be a public relations disaster for China, it would almost certainly move the rest of the world to act—as it did in Yugoslavia—for humanitarian reasons, in spite of the considerations of sovereignty. This would be particularly true if the conflict became protracted. The likelihood of such protraction, underlined by the Kosovo experience, is perhaps the single most worrying aspect of the whole thing for China.

Financial markets would also be harmed. It goes without saying that Taiwan would suffer severe losses, but so, too, would China. When President Lee gave his July 1999 interview regarding statehood, Taipei markets fell—but so did Hong Kong's and Shanghai's, and the same happened in early 2000 when Beijing again turned up the temperature. China's economy is already on a downward trajectory and needs all the foreign investment it can get. Hostilities around Taiwan would almost certainly stop the flow—not only from Taiwan itself, but also from Japan and Europe, and probably from Hong Kong as well, because, as markets there fell and trade stopped, few would be in the mood to invest.

Furthermore, and this should be stressed, a Kosovo-style attack on Taiwan would undoubtedly close U.S. markets completely to Chinese exports. Nothing from China would be permitted into the country. Given China's great dependence on foreign sales and foreign currency, this would be a disaster.

International Aspects.

If one envisions a PRC attack on Taiwan as looking like Kosovo, then it is hard to imagine any way that the rest of the world would stay out. We are talking about the heart of maritime Asia, as well as major sea-lanes, where billions of dollars of commerce pass daily. The immediate pressure would be on China to stop—and once stopped, the Chinese attack, like the NATO bombing, would be difficult to restart. Having initiated war and then been forced to back down, China would be in a weaker negotiating position than at the outset. An attack on Taiwan would click the lock on a circle of alliances that would effectively contain China, with Japan almost certainly joining.

What of the doomsday scenarios: Chinese nuclear threats against Japan or the United States, cruise missile attacks on U.S. carriers, neutron bomb "ethnic cleansing" of

Taiwan that would leave the valuable infrastructure intact to be inhabited by a new, more tractable population?

War does, of course, bring disaster. But disasters come from desperation, not calculation, and the operation we are hypothesizing would be, like Secretary Albright's war, intended to be strictly limited. When such a limited operation failed in Kosovo, NATO escalated, but against an adversary that was itself incapable of escalating and that had no ally willing to run the risk on its behalf. To escalate against Taiwan, however, would invite desperate retaliation from the island, while to attack U.S. forces would guarantee a U.S. response.

The states most likely to use weapons of mass destruction are not the great powers who stand to lose a great deal in any war, but rather those countries or regimes for whom defeat can mean destruction or collapse—Israel, North Korea, Iraq, Pakistan, and so forth—to which one could add Taiwan and South Korea, were they nuclear powers. RMA, however, may render nuclear explosions obsolete, given the kind of damage that new-style conventional weapons can do.

Taiwan in particular is developing non-nuclear weapons systems that can nevertheless do substantial damage to China. Given the political weakness of the Chinese regime and the need it would have for a quick and costless victory, the complications such weapons would introduce could well undo the entire strategy—even if the rest of the world did nothing.

Chinese Responses.

China should have learned a great deal about its own diplomatic and military weakness in the course of the Kosovo war. Traditional Chinese strategy stresses splitting alliances and playing one power off against another. But in spite of Chinese efforts to lobby "soft" NATO countries such as Italy, the Western alliance stayed united to the end of the campaign. Nor did the hope China placed on Russia materialize. Moscow stood aside even though it could very easily have disrupted NATO's plans. Finally, the rest of the world, as represented in the United Nations, proved a weak reed for China as well. On May 13, 1999, China failed to get the Security Council to "deplore" the bombing.[17] These results must have been disappointing, given the stress China lays on her influence in international institutions and the power of her veto.

Rhetoric at the time suggested that frustration over failed diplomacy turned into anti-Americanism in China. The United States emerged as the spider spinning the web in which China itself feels increasingly entangled—a web of economic dependence, of international norms, and increasingly of sanctions, utterly at odds with the Chinese government's policies—though perhaps not unwelcome by a lot of China's people. But almost a year later another explanation is also beginning to emerge. This is that the debate over Kosovo fits into and exacerbates an already existing struggle within the Chinese government in which domestic hard-liners such as Jiang Zemin are increasingly positioning themselves beside the international hard-liners in the military to attempt to master the ever more volatile domestic situation.

Certainly some of the rhetoric sounded as if it was intended for domestic political consumption, not for analytical purposes. One article predicted that NATO would spread into Asia, and that "the first Asian state to join NATO could be Taiwan" to be followed by Xinjiang, at the urging of Turkey.[18] But who can doubt that the Turks of Xinjiang are increasingly restive, well-armed, and in touch with the rest of the Islamic and Turkish world of Central Asia, and that Chinese attempts to woo Kazakhstan have not proved entirely successful? The problem is a real one, and at the moment Beijing has no answer for it except threats and repression—which failed in the past when the Turks were weaker, the Chinese stronger, and the international community in the dark.

Likewise with Taiwan. It probably will not join NATO. But if one were asked to choose the more probable course of events, given China's escalating threats: do the United States and Japan cut Taiwan off and force it to sign on the dotted line? Or do they gradually increase their ties, providing key diplomatic support and missile defense—with the effect of frustrating China's efforts—then I think the second is far more likely, as China understands.

The problem with Tibet is similar. It cannot be suppressed. The Chinese plan to choose and educate the next Dalai Lama was unworkable even before the current Dalai Lama dispatched it by stating that he would not be reincarnated as a Tibetan under Chinese control. The kidnapping by China of the Panchen Lama and the flight to India of the Karmapa Lama further underline the impossibility of Chinese plans. India is less strong militarily than China but also less overextended, and its presence to the south and sympathy for Tibet guarantees that this issue will not disappear.

Nor can China be certain that the United States might attempt to coerce China as it did Kosovo with high-technology weaponry. North Korea certainly fears that it may be a target. Pyongyang watched the war far more closely than it did the Gulf War, and has evidently expressed fear to Beijing that it could be the next Yugoslavia—and it could be.[19]

In fact, Western use of force against China proper would probably be as ineffective as the Kosovo operation and a good deal messier. China is not good at force projection, and not only because she lacks the hardware. Force projection goes against traditional approaches to war, which lay stress on using the opponent's own weight to bring him down. Western force can halt Chinese attacks outward anywhere in Asia—but it cannot bring Beijing to terms. If it did not escalate to the nuclear level, a possibility not to be excluded, a Kosovo-style operation against China would most likely simply go on and on, as China absorbed the punishment and occasionally struck back, hoping to get lucky one day and kill enough Americans to force an agonizing reappraisal on Washington. That was how China dealt with Japan half a century ago.

Still the question arises for Beijing, Pyongyang, and the others of how the United States is to be kept out? Nuclear threats seem one possibility. Wang Xiaodong, author of a recent article on nationalism and democracy, believes that building precision weapons (as some in the PLA undoubtedly want) makes no sense. He says:

> We must strengthen our nuclear capability . . . We need much more weapons of mass destruction which can reach cities in North America. It is important to remember that the key difference between China and Serbia . . . is that we have the nuclear bomb.[20]

That sounds plausible until one considers that the Soviet Union had many more bombs than the United States, and that didn't save it. Furthermore, the sort of buildup of nuclear weapons in which China is currently involved will certainly elicit a countervailing response, even if not immediately. Japan, Korea, Taiwan, Vietnam, and other Asian countries will join India in the recognition that only their own deterrent can ultimately deter a threatening China. American attempts to discourage them from developing such capabilities will only make the situation worse. But an Asia containing several nuclear powers with advanced missiles and other forces comparable to what China is developing now can only hurt China's interests. The United States operates with other powers by allying with them. China at present seems to prefer to threaten them.

Conclusion.

The trends that have so worried China look likely to continue. Democratization is a risky process, but it proceeds apace in Asia, where Indonesia most recently has chosen that pathway—and not coincidentally, granted independence to East Timor, a major step toward ending conflict and restoring peace in the area. In China itself, problems are now of such complexity that it is hard to imagine the autocratic government will be able to deal with them. Few of these problems have anything to do with foreign policy: the genuinely big issues in China involve such things as unemployment, environmental degradation, the need for law and democracy, and the rights of citizens. Nor can the government any longer simply pretend, as it did in the past, to have solved these problems. Penetration of international news and the rising educational level have decisively undermined the effectiveness of central propaganda. Meanwhile the willingness of hundreds of members of the China Democratic Party to go to prison rather than recant, and the continued failure of Beijing's much-ballyhooed campaign to "crush" the Falun Gong organization, further demonstrate the steady erosion of the center's power. Xinjiang shows no sign of calming down despite a massive military presence, and Tibet is similarly deadlocked. These problems show how the great currents of contemporary history are affecting China. Nor is military might an answer. It cannot solve domestic problems, and, as Kosovo shows, it is almost equally helpless against external problems.

Beijing continues to talk about a decisive conquest of Taiwan by means of missiles and smart bombs, so rapid that Washington would have no time to act even if it chose to. But the reality, as China must realize, is that such an onslaught would be indecisive and quickly escalate.

Kosovo shows just how appealing such fantasies continue to be to states that should know better. That force is rarely as effective as hoped for and often escalates out of control is a lesson that Europe and the United States should have learned in the 20th century. Far better that China should learn it by studying Kosovo than by experimenting with Taiwan.

CHAPTER 7 - ENDNOTES

1. "China Slams NATO Strikes in Yugoslavia, Calls for End of Action," *AFP* Beijing, April 15, 1999.

2. *People's Daily*, June 22, 1999, p. 6.

3. "China Reiterates Stand on Kosovo Issue," *Xinhua* Beijing, May 6, 1999.

4. "'Ethnic Cleansing' Conceded," *South China Morning Post*, April 22, 1999.

5. *China Economic Times*, April 20, 1999.

6. *Ibid.*

7. "Milosevic Hailed for Battling 'Hitler' Clinton," *South China Morning Post*, April 5, 1999.

8. *Ibid.*

9. *Ibid.*

10. "Mainland's Flip-flop Over the Balkans," *South China Morning Post,* April 1999.

11. Willy Wo-lap Lam, "Party Warns of US Expansionism," *South China Morning Post,* March 18, 1999.

12. "Evolving Democracy," *New York Times,* July 19, 1999.

13. "The Heart of Chinese Sovereignty," *South China Morning Post*, June 12, 1999.

14. Bryan Bender, "Kosovo Lessons Won't Mean Big Changes by USA," *Jane's Defence Weekly,* July 14, 1999, p. 3.

15. *Ibid.*

16. Ainslee T. Embree, ed., *Encyclopedia of Asian History*, Vol. 2, New York: Charles Scribners' Sons, 1988, p. 223.

17. Willy Wo-lap Lam, "Beijing Targets 'Soft' NATO Nations," *South China Morning Post,* May 14, 1997.

18. Article by Wei Mufu and other researchers at the Asia Pacific Council, "The Kosovo Affair and Asia-Pacific Security," quoted in "The Heart of Chinese Sovereignty," *South China Morning Post,* June 12, 1999.

19. Willy Wo-lap Lam, "Fears North Korea NATO's Next Target," *South China Morning Post,* May 28, 1999; "N. Korea Active in Reporting Kosovo Conflict: Defense Min." *Korea Times,* April 12, 1999.

20. Mufu.

CHAPTER 8

CHINA'S MARITIME STRATEGY

Bernard D. Cole

Introduction.

The Asia-Pacific is an overwhelmingly maritime region, defined by the Pacific and Indian Oceans; and by the Yellow, East China, and South China Seas, which border the nation for which they are named. The state of maritime strategic thought in the People's Republic of China (PRC) today, and the likely development of that thought over the course of the next decade, are open questions. Their answer requires, first, a review of the geo-political context facing Chinese maritime strategists. Second, a discussion of the development of maritime strategy in the PRC will be followed by a brief review of specific maritime strategic missions and capabilities of the People's Liberation Army Navy (PLAN). Finally, future prospects for the navy's strategic role will be surveyed.

In 1953, Mao Zedong wrote "we must build a strong navy for the purpose of fighting against imperialist aggression." In 1979, Deng Xiaoping called for "a strong navy with modern combat capability," and in 1997, Jiang Zemin urged the navy to "build up the nation's maritime Great Wall."[1] China is clearly expanding and modernizing its navy, about to celebrate its fiftieth year, by focusing both on the hardware and strategy required for a modern fleet.

The PRC has not been reluctant to use military force, including warships, to resolve international disputes—more than 16 times, in fact, since 1949.[2] A major goal of China's naval modernization is to change the PLAN from a coastal, "brown-water" force to an open ocean "blue water" force.[3] "Blue water" is part of a maritime strategy paradigm that includes "brown" and "green" water. Brown water refers to littoral ocean areas, within about 100 nautical miles (nm)[4] of the coastline; green water is less definite, referring to ocean areas from about 100 nm to the next significant land formation. For China, green water extends from brown water to Okinawa or from brown water throughout the South China Sea (SCS). "Blue water" is represented by General Liu Huaqing's "second island chain," delineated by a line from Japan through the Bonin Islands, the Mariana Islands, and the Caroline Islands. A "blue-water" PLAN, then, will have to be capable of controlling events to a distance of at least 1,500 nm from China's coast, including the Yellow, East China, and South China Seas.

The classic maritime strategic phrase for such control is "Command of the Sea," most simply defined as the ability to use the sea while denying its use to an adversary. "Sea Control" is a lesser but nonetheless powerful concept, defined as a nation's ability to "command" a discrete ocean area for a limited period of time—sufficient to achieve limited strategic goals.

The counterpoint to sea control is "Sea Denial": denying an adversary use of a discrete maritime area without using it oneself. All of these concepts require a nation to be effective in the air, as well as on and beneath the ocean surface. Sea denial in littoral waters is a particularly attractive and inexpensive option for even a small naval power, if it has access to mines, missiles, small surface ships and submarines, and shore-based aircraft—as does China.[5]

China is learning that modern navies are technology-dependent and resource-intensive; they cannot be built (or even bought) quickly.[6] Indeed, as the noted British strategist Julian Corbett implied, "modern naval strategy may be the application of professional experience to the solution of technical problems."[7]

Several factors influence modern naval strategy making. These include:

1. Training and education programs; professional specialization of the officer corps;

2. naval systems and platforms costs, capabilities, and sustainability;

3. the national scientific and industrial infrastructure for research, development, and production of naval warfare technology and systems;

4. the ability to derive doctrine and tactics;

5. the ability to administer, operate, and command and control tactical units beyond individual ships;

6. sources of intelligence, and its production, analysis, and dissemination;

7. service-wide naval strategic planning;

8. national naval leadership; and,

9. the place held by naval strategists in the national strategy-making structure.[8]

We will return to these points to evaluate China's development of a maritime strategy.

Maritime strategy should reflect Colin Gray's dictum that "man lives on the land, not on the sea, and conflict at sea has strategic meaning only with reference to what its outcome enables, or implies, for the course of events on land."[9] Gray cautions, however, that in all the history of war, "the enemy who is confined to a land strategy is in the end defeated."[10] China historically has viewed the sea as an invasion route by foreign aggressors, rather than as a medium for achieving national goals. This attitude appears to have changed during the past decade and a half, as China's view of post-Cold War East Asia has focused on off-shore sovereignty, economic, and resource issues.

Adopting a maritime strategy requires naval leadership with a strong voice in the always-contentious national and military resource allocation process. Also relevant is the recent development in maritime strategy, epitomized by U.S. adoption of the "From the Sea . . ." Series of strategy documents, that sea control, *per se*, while still important, is being matched in significance by the ability to project naval power ashore, to directly affect events on the land. This development will heighten, not lessen, the importance of sea power.

Strategy, Politics, and Geography.

China historically has been a continental rather than a maritime power, despite possessing over 11,000 miles of coastline and more than 6,000 islands. China has only episodically built and employed a strong navy, most recently over a century ago, during the late 19th century "Self-Strengthening Movement."

The Chinese currently view the international security situation in Asia as peaceful but dangerous. President Jiang Zemin was recently quoted as stating that the world "is moving deeper towards . . . . the relaxation of the international situation and world peace," but that "the Cold War mentality still lingers on as hegemonism and power politics manifest themselves . . . . New forms of 'gunboat policy' are rampant."[11]

From Beijing's point of view, U.S. naval dominance in East Asia seems assured in the near term. In the longer term, however, the American presence is unclear and Japan is a military threat. The maritime environment also directly impacts China's most serious resource problems: energy and food, both of which depend on the sea.[12]

Since the U.S.-led NATO war against Serbia, the Chinese press has been marked by strong condemnation of the "new gunboat policy" the United States is accused of pursuing, "under the cover of . . . so-called security globalization," a plan "to dominate the world."[13] In the Asia-Pacific region, the United States is charged with using the new defense agreement with Japan and theater missile defense (TMD) "to prepare the ground for future military intervention," and with a "Eurasian strategy" to position the United States for a "one superpower-dominated 'U.S. century'."[14]

China insists it is a developing country still facing injustice and unfair treatment on the international scene.[15] General Zhang Wannian, Vice-Chairman of the Central Military Commission (CMC), reportedly has stated that China is "threatened by hegemonism and power politics, by militarism, and by foreign military intervention in Taiwan. ... [We] must absolutely not lower our guard."[16] He accuses the United States of seeking the collapse of the Beijing regime during a "full-fledged civil war," as a result of which China "will disintegrate."[17]

This is a strategic picture of China besieged by a hostile world in which the United States pushes the west and a militaristic Japan to contain China and prevent it from attaining its rightful stature as a world power. What maritime strategy will serve China's national security concerns—especially in view of U.S. determination to place "humanitarian rights" above national boundaries?[18]

The Navy and China's Defense Modernization.

The current Chinese campaign to modernize its military follows the "sea change" in strategic thinking that occurred in 1985, when expectations of global nuclear war or large-scale conflict with the Soviet Union gave way to a focus on small, local wars on China's periphery. The local, limited wars envisioned include (1) small-scale conflicts in disputed border areas; (2) conflicts over disputed islands or ocean areas; (3) surprise air attacks; (4) deliberate incursions into China; and (5) counterattacks by China against an aggressor or "to uphold justice and dispel threats."[19]

This was an extremely important shift for Chinese maritime strategic thought. It moved the navy from army acolyte to prominent participant in possible operational scenarios, including threats from Japan, Taiwan, India, and contentious maritime claims.

Following this strategic change, China reduced the number of military personnel, but to **increase** rather than **decrease** military power, as the nation seeks to fund state-of-the-art technology at the cost of low-skilled ground troops. The PLAN particularly benefits from this phenomenon, since compared to much army equipment, the systems required by naval forces are technologically intense, require long periods of research and development, and demand operating personnel who are intellectually capable and the product of extensive training. Furthermore, advanced naval systems require large investments in maintenance and logistical support.

The PLAN was also able to continue older programs, including the development of nuclear-powered submarines and the expansion of aviation forces to gain control of the air over ocean areas. The marine corps, first formed in 1954 but disestablished in 1957, was reorganized in 1980 for additional amphibious assault capability.

The Evolution of Maritime Strategy in China.

China has always focused on continental security concerns, but has not ignored its maritime boundaries. Witness the early 15th century voyages of Zheng He to the east coast of Africa and the Persian Gulf. These cruises represented a standard of Chinese shipbuilding, voyage management, and navigation ability well beyond European capabilities. Zheng He led a large fleet of ships, the largest displacing over 400 tons, half-way around the world at a time when Portuguese explorers were still feeling their way down the west coast of Africa in 50-ton caravels.

The Ming rulers deliberately ended these voyages for domestic financial and policy reasons, just at the time when the European nations were seizing upon them as a means to achieve economic wealth and to proselytize. By the late 18th century, the Europeans were the maritime powers of the world, and China their victim.[20]

China under much of the Qing Dynasty was an Asian maritime power, in terms of oceanic trade and fishing, but did not build a navy capable of defending the nation against maritime interlopers. Late 19th century efforts to correct this shortcoming lacked both a coherent strategic rationale and adequate support from the national government; the nascent Chinese fleet came to grief in defeats by France in 1884 and Japan in 1895.

The regimes of the first half of the 20th century developed a navy almost entirely through seizing former Qing warships or obtaining them from foreign nations. No distinct maritime strategy was developed.

The Early Years: 1949-1954. The communist victory in 1949 was an army victory; there was no navy arm of the PLA. The Chinese Communist Party (CCP) did believe, however, that China's 19th and 20th century humiliation had been greatly facilitated by the imperialists' ability to invade from the sea. The new government in Beijing understood the need to defend its coastline and island territories, and considered itself threatened by two would-be aggressors: the truncated *Kuomintang* (KMT) regime that had fled to Taiwan, and the United States.

The PLA was unable to project power across even the relatively narrow (90-110 nm) Taiwan Strait in 1949 to complete the defeat of the KMT forces. Although expelled from the mainland, Chiang Kai-shek still possessed a navy effective enough to stop the PLA at the water's edge. Hence, the new Chinese regime in 1949 immediately moved to create a navy to continue the fight against Chiang's forces and to defend the nation against "imperialist aggression from the sea."[21] A new Chinese navy was also needed to establish law and order on coastal and riverine waters, to combat the KMT blockade, to help the army capture the off-shore islands still occupied by the KMT, and to prepare for the capture of Taiwan. The "East China Navy" was organized in 1949, and the PLAN was officially established in May 1950.[22]

The Taiwan government continued to conduct hostile operations against the mainland, raiding coastal installations, landing agents, attacking mainland merchant craft and fishing vessels, and threatening invasion on a larger scale. China needed a defensive navy, one that would be relatively inexpensive to build, and could be quickly manned and trained.[23]

The new PLAN, established with Soviet assistance and guidance, consisted for the most part of small patrol craft suitable to combat the coastal threat from Taiwan. The Soviet military assistance obtained by Mao Zedong during his 1949-50 visit to Moscow included naval equipment and advisors, who instilled the Soviet "Young School" of maritime strategy. The Young School, developed in the Soviet Union shortly after World War I, was based on conditions particular to post-revolutionary Russia:

1. a new regime that was under military and political attack by several capitalist countries and had not completely quelled domestic fighting;

2. furthermore, a regime that *expected* to be besieged and attacked by capitalist nations, with amphibious invasion a current fact and future threat, especially from "the ultimate bastion of imperialism, the United States."[24]

3. a navy that was in disarray;

4. budgetary shortages that limited the amount available to spend on expensive naval systems;

5. lack of an industrial infrastructure to indigenously produce modern naval armaments; and,

6. a maritime frontier that was hemmed in by adversarial fleets and bases.

These conditions applied to China in 1949, as well, as did the additional problems of no recent maritime tradition or trained naval force.

The Soviet Union sent an initial cadre of 500 naval advisors to China in 1950, which grew to between 1,500 and 2,000 by 1953. These advisors paralleled the Chinese chain-of-command from Beijing headquarters to individual ships and squadrons, thus inculcating Soviet naval doctrine in the new navy. "Large numbers" of Chinese officers, including the new head of the PLAN, General Xiao Jingguang, received training in the Soviet Union.[25] Soviet assistance included four old submarines, two destroyers, and a large number of patrol boats. This unimpressive force also included about 10 corvettes, 40 ex-U.S. landing craft, and several dozen miscellaneous river gunboats, minesweepers, and yard craft, all seized from the Nationalists.[26]

A large shore-based infrastructure, including shipyards, naval colleges, and extensive coastal fortifications, was also built with Soviet help.[27] The high point of the cooperative period with the Soviets was the February 1959 "New Technology Agreement," which provided for the sale to China of additional conventionally powered submarines, torpedo boats, and missile boats.[28] Although its maritime strategy in the early 1950s was primarily defensive, Beijing worked to develop the offensive capability to recover the many offshore islands still occupied by the KMT, which was to culminate in the conquest of Taiwan in August 1950.[29]

When President Harry Truman ordered the U.S. Seventh Fleet into the Taiwan Strait in June 1950 at the outset of the Korean War, he explained America's reentry into the Chinese civil war as a means of preventing either side from attacking the other. Actually, he was committing the United States to the defense of Taiwan—after having refused to do so for many months.[30]

President Dwight Eisenhower announced in February 1953 that the U.S. fleet would no longer interfere in the strait, thus in theory "unleashing" Nationalist forces on

Taiwan to attack China. In December 1953, Mao Zedong assigned the PLAN three priority missions: (1) eliminate KMT naval interference and ensure safe navigation; (2) prepare to recover Taiwan; and (3) oppose aggression from the sea.

The Chinese and Taiwanese forces resumed amphibious attacks and counter-attacks, which ended in Chinese possession of all the significant offshore islands except Quemoy, Matsu, and, of course, Taiwan. The PLA also succeeded in stopping most of the attacks on merchant and fishing vessels, as well as the Taiwanese raids on the mainland. The PLAN had been organized, sent to sea, and proven effective as a Soviet-style coastal defense force, while adhering to the rubric of "People's War."

The Years of 1955-1959. The Korean War experience must have presented mixed lessons to the Chinese Navy. The amphibious landing at Inchon in September 1950 was a major turning point of the war, while allied command of the sea allowed free employment of aircraft carriers and battleships to bombard Chinese and North Korean forces. It was not significantly a maritime conflict, however, and the PLA may have thought in 1954 it had demonstrated the ability to campaign successfully without maritime support.

Beijing thought the American invasion threat could be countered by relatively short-range defensive sea forces; no ocean-going Chinese navy was planned after 1954. During the latter part of the decade, after China's commanders had witnessed first hand in Korea the effects of modern weaponry, and had been threatened with nuclear warfare, some PLA leaders wanted to modify Mao's theory of "People's War" to one of "People's War Under Modern Conditions." Their ability to build a PLA on this principle, however, was limited by the necessity to continue conforming to Maoist ideology.

The heart of China's national security strategy continued to be one of "active defense": the enemy would be lured inside China's borders while the nation mobilized. A strategic offensive would then be launched to drive the invader from China. The navy tried to modernize, with an emphasis on technology and technical training, while adhering to a brown water defensive strategy.

Attempts in the 1950s to develop a specific and perhaps independent strategic role for the PLA fell victim to the triumph of politics over technology. One prominent victim was Minister of Defense Marshal Peng Dehuai, who stressed the importance of scientific and technological progress and supposedly opined, with respect to the navy, that "People's War and such stuff are outdated [at sea]."[31] Throughout the ideological turmoil of the late 1950s and the 1960s, Beijing did invest heavily in a determined effort to develop nuclear weapons, missiles, and the nuclear-powered submarines from which they could be launched.[32]

A New Situation: 1960-1976. By 1960, the PLAN was considerably larger than in 1955, but it had made limited strides towards modernization. The force centered around eight ex-Soviet destroyers, two dozen old Soviet submarines, and a motley collection of foreign-built patrol boats, landing craft, and yard craft. The only vessels built in China were some of the patrol boats.[33]

The 1960s were marked by major foreign and domestic events that further constrained Chinese interest in developing a sea-going navy and attendant strategy. Most important was the split with the Soviet Union, signaled during Nikita Krushchev's October 1959 meeting with Mao Zedong in Beijing, and dramatically executed in mid-1960, when Krushchev withdrew Soviet advisors (and their plans) from China. Beijing's massive civilian and military development projects were left in turmoil. Other significant events in the early 1960s included war with India, the reemerging Vietnam conflict, turmoil in the newly emerging African states, and would-be revolutionary movements throughout Southeast Asia.

These factors continued to limit PLAN modernization in the 1960s. Marshal Lin Biao, as minister of defense, attempted to follow a policy of technological development with "politics in command." The decade ended, however, with Lin coming down solidly on the side of "politics," writing "long live the victory of people's war."[34]

The massive Soviet threat and the PLA's lack of mobility drove China's national security strategy: very large ground forces were needed. Mao's concept of People's War remained the guidance for the small navy as well as the other branches of the PLA, as did continued adherence to the Young School—modified by the development of nuclear-powered ballistic missile submarines.[35]

America's involvement in Vietnam and Taiwan's failure to act on the rhetoric of invading the mainland meant that China faced no overseas threat.[36] By the end of the decade, however, relations with the Soviet Union had deteriorated to the point of armed conflict along the Amur River. The former ally was now the enemy. Soon the erstwhile enemy, the United States, would be China's strategic ally.

Internally, the Great Proletarian Cultural Revolution (GPCR) made impossible any shift from Maoist theology; hence, the navy continued to follow a strategy of coastal defense. This meant serving as an extension of the army, with little modernization: People's War held that technology and weaponry were insignificant compared to the effect of revolutionary soldiers imbued with Mao's ideology. In May 1975, however, at a meeting of the CMC, Mao Zedong reportedly directed the development of a modern navy.[37] The GPCR seriously hampered technological development, even in the relatively sacrosanct missile, submarine, and nuclear weapon development programs. PLAN development was probably retarded by a decade or more, because of the program restrictions and personnel losses that resulted from the political maelstrom.

Even at the end of the GPCR, PLAN modernization was hamstrung by the "Gang of Four." Mao's widow, Jiang Qing, led the attack on naval missile development. Another member, Zhang Chunqiao, expressed the Gang's anti-navy position and support for the "continentalist view."[38] By 1970, however, the PLAN had moved into the missile age, deploying a Soviet-designed ballistic missile submarine and 10 Soviet-built patrol boats armed with cruise missiles.[39]

After the GPCR. As the 1970s progressed PLAN strategic missions still fell under the Young School: assistance to the army; off-shore patrol against criminal activities such as smuggling, piracy, and illegal immigration; life-saving; and safety of navigation. Beijing also perceived its ancient antagonist, Japan, reemerging as a military force to be reckoned with.

Meanwhile, the Soviet Navy in the 1960s and 1970s had undergone a dramatic change under the leadership of Admiral Sergei Gorshkov, partly from the impetus of the Cuban Missile Crisis's demonstration of Soviet maritime weakness. Under Gorshkov's guidance, the Soviets attempted to build a worldwide fleet to match that of the United States. This fleet's missions in time of war were:

1. defense of off-shore areas; countering an adversary's strategic strike systems;

2. sea control in Fleet Ballistic Missile (FBM) Submarine operating areas;

3. strategic nuclear strike;

4. disrupting an adversary's sea lines of communications (SLOCs); and,

5. protecting friendly SLOCs.

Gorshkov's maritime strategy also included specific peacetime tasks: (1) showing the flag; (2) gaining international respect; (3) supporting economic interests; (4) managing crises; (5) limiting an adversary's options; (6) exercising local sea control; and, (7) use in local wars.[40]

A similar naval revolution did not occur in China in the 1960s or 1970s, but when the Soviet Union demonstrated its new global naval power in the 1975 *Okean* exercises, Beijing strategists moved the Soviets to the top of the list of China's maritime threats. The strategy of People's War was deemed insufficient. Maritime strategic thought was reinvigorated, as Chinese planners thought about extending naval power beyond the immediate coastal arena against potential Soviet actions.

Concern about Soviet aggression rose in concert with continued determination to ensure the viability of Beijing's territorial claims throughout East Asia. Taiwan was the most important of these, but China also had claims in the East and South China Seas. In 1974 a brief but effective action against South Vietnamese naval forces in the SCS resulted in Chinese possession of the disputed Paracel Islands.

Beijing ended the 1970s without a clear maritime strategy. Fear of the Soviet Union provided impetus to the development of a seaborne nuclear deterrent force, with Mao declaring that the navy had to be built up "to make it dreadful to the enemy."[41] The PLAN for the first time was led by Chinese-built warships: *Luda*-class guided-missile destroyers, frigates, and fast attack missile boats. The submarine force included the first Chinese-built nuclear-powered attack submarine, as well as about 60 conventionally powered boats. The 1970s saw the deployment of the first PLAN that was primarily Chinese-built, although still heavily dependent on Soviet designs.[42]

At the end of the decade, however, Deng Xiaoping reemphasized the navy's role as a coastal defense force, a view retained throughout the first half of the 1980s. At that point, the navy came to be viewed as something more than an adjunct to the ground forces. Maritime power was emerging as an important instrument of national security strategy.

PLAN expansion and modernization were spurred by the coastal concentration of China's burgeoning economy and many military facilities. Furthermore, the resources necessary for a modernized PLAN became available as a result of China's dramatic economic development and increasing wealth. The navy's strategy began developing beyond coastal defense.

China's widening maritime interests and increased budget resources after 1979 did give rise to increased interest in a strong navy of large modern warships, similar to that championed by Gorshkov. If built, the new force would have to be capable of projecting power throughout the area Beijing considered important to its national security—all of East Asia—and to execute the post-1985 strategic rationale from massive continental war to regional wars on China's periphery.

Liu Huaqing's Vision. The chief architect of China's emerging maritime strategy in the 1980s was General Liu Huaqing, commander of the PLAN from 1982 to 1987 and then Vice-Chairman of the CMC to September 1997. Liu advanced a maritime strategy designed to address China's national security concerns, and one that would thrive in the army-dominated PLA.[43]

Liu wanted to change the PLAN's national mission from coastal defense to "offshore active defense." This strategy was described as including: (1) stubborn defense near the shore; (2) mobile warfare at sea; and, (3) surprise guerrilla-like attacks at sea.[44] He delineated two strategic maritime areas the nation should be capable of controlling. The first of these, under Stage One of Liu's strategy, includes the Yellow Sea, facing Korea and Japan; the western East China Sea (ECS), including Taiwan; and the SCS.[45] These fall within China's defined area of vital national interests: territorial claims, natural resources, and coastal defense. The area is also delineated by the "first island chain," a line north to south from the Aleutians through the Kuriles, Japan, the Ryukyus, Taiwan, the Philippines, and Indonesia. The goal for establishing Chinese control of this maritime area was 2000.

The second strategic maritime area, under Stage Two of Liu's strategy, is delineated by the "second island chain,"a north-south line from the Kuriles through Japan, the Bonins, the Marianas, and the Carolines, and would mean Beijing's control of all of East Asia's vast ocean areas, nominally by 2020.[46] China's ability to control this area would demand very significant national resources for its navy and air force. It would also require, first, that the United States withdraw its military presence from the region. Second, Japan would have to sit idly by in the face of U.S. withdrawal, and not engage in a maritime and aerospace arms race.

Ironically, defining "stages" of maritime theaters by fixed geographic boundaries reveals a strong continentalist perspective, even in the mind of China's most prominent post-1949 admiral. It violates the argument of western maritime strategists that while the soldier thinks of terrain and theaters, the sailor of necessity thinks in wider terms, outside immediate physical limits—there is no "terrain" at sea.[47]

The third stage of Liu's maritime strategy describes the PLAN as a global force by 2050. A step towards this goal occurred in the spring of 1997, when within a 3-month period China deployed multi-ship task groups composed of warships and logistics support vessels, one to Southeast Asia and a second to North, Central, and South America. This was the widest-ranging Chinese naval deployment since the voyages of Zheng He and was a significant accomplishment.

Scenarios. Liu's strategy outlines a possible direction for future modernization and growth. It delineates control of vast oceanic expanses; however, a very difficult task simply by virtue of the geography, not to mention other nations that would object to Chinese hegemony over such a large portion of the earth's surface. Of course, the PLAN is currently incapable of executing its missions within even the "First Island Chain," although the target year will soon be here.

Recent deployments to the Western Hemisphere were just an initial step; there currently is no hard evidence that China's national leadership has decided to shift its budget priorities to the extent necessary for realization of Liu Huaqing's Phase II by 2020.[48]

Liu Huaqing is no fool and one suspects that this three-phase strategic progression was designed primarily for domestic consumption, to win resources for the PLAN. Liu would have been following a path not unlike that pursued by naval expansionists in turn-of-the-century America and Germany as they built modern navies in conjunction with their nations' rapidly expanding economies. Liu directed study and elaboration of his strategic concepts, which emphasized naval missions well to seaward of the coastal zone that has formed the basis for past PRC maritime strategy.[49] PLAN modernization is not viable without a well-articulated offshore mission, supporting China as the strongest maritime power in East Asia and as a major power in the Pacific.

The PLAN has not been tasked with all of the "new" Soviet maritime strategic objectives of the 1970s. It has no "FBM operating areas" to control, for instance, and its "strategic nuclear strike" capability is almost non-existent—but both of these missions will require support if China succeeds in building and deploying three or more new (Type 094) ballistic missile submarines.[50]

The PLAN has, however, adopted peacetime strategic missions almost identical to those outlined by Gorshkov—who cited the American threat as his basic justification for a strong navy. Writing in 1975, he accused the United States of following an "oceanic strategy" of aggression against the Soviet Union.[51] Chinese strategists today use similar words; the Soviet-Russian maritime strategic influence remains strong in China's navy.[52]

Maritime Strategic Interests.

Jiang Zemin has been quoted as pledging the PLAN "to safeguard the sovereignty of China's territorial waters, uphold the country's unity and social stability and create a safe and stable environment for the nation's economic development."[53] The PLAN's primary mission is defense of the homeland, a formidable task, given China's long sea frontier from Korea south to the Indochinese Peninsula. This great maritime sweep is marked by major offshore island chains from Japan to the Philippines, New Guinea, and Indonesia, as well as by numerous lesser islands.

The PLAN's ambition to play a central national security role is evidenced in the 1991 statement by then PLAN vice-chief, Vice Admiral Cheng Mingshang, that the navy is

> the tool of the state's foreign policy . . . . an international navy can project its presence far away from home. It can even appear at the sea close to the coastal lines of the target countries . . . . This has made the navy the most active strategic force in peacetime, a pillar for the country's foreign policy and the embodiment of the country's will and power.[54]

Economic justification for a strong Chinese navy rests on the concentration of modern economic interests and growth in the special development zones clustered along China's seaboard.[55] Seabed minerals, especially potential petroleum deposits in the SCS, are also important, and there is evidence of significant deposits elsewhere on the continental shelf, including around the Diaoyutai between Taiwan and Okinawa, also claimed by Japan as the Senkaku Islands.

China's most urgent international security concerns—Korea, Japan, Taiwan, the South China Sea—lie to seaward and have a strong maritime flavor. The national civilian and military leadership must decide how to allocate defense resources among the various services to deal best with these concerns—and the associated maritime sovereignty issues.

Additional missions for the Chinese Navy include anti-smuggling, anti-piracy, fisheries protection, and defense of SLOCs.[56] This last mission is supported by the fact that as much as 50 percent of China's economy depends on foreign trade, about 90 percent of which is transported in maritime shipping. China's large and growing merchant fleet calls at over 600 ports in more than 150 countries.[57]

Northeast Asia. China's North Sea Fleet is responsible for the area from the Korean Border (the Yalu River) to about 35°-10'N, corresponding to Shenyang, Beijing, and Jinan Military Regions. Its primary bases are at Yuchi, Chengshan, Weihai, Huludao, Lushun, and its headquarters, Qingdao. Its force structure includes two submarine, three escort, one mine warfare, and one amphibious squadrons.[58] This fleet faces a complex maritime situation: three other Northeast Asian nations—Russia, South Korea, and Japan—already possess capable navies, although Russia's fleet is only a shadow of its former self.

Korea is particularly sensitive because North Korea may be armed with nuclear weapons. Beijing's relationship with the peninsula is also complicated by the ethnic Korean population in China's northeastern provinces.

Japan is not as pressing an issue, but in terms of China's national strategic concerns looms very large on the horizon of the next century. Ancient disputes and rancor combined with World War II grievances and suspicion of future Japanese aggression create an edgy relationship. It is inherently a maritime relationship, given the seas that lie between the two nations act as a natural barrier to any but seaborne or airborne interaction.

Taiwan. Taiwan of course is the essence of Beijing's strategic concerns. Despite concern about the United States, China refuses to renounce the use of military force to ensure the reunification of Taiwan. Beijing must count on the PLAN for policy options ranging from intimidation to outright invasion.[59]

The East Sea Fleet presumably has local planning responsibility for contingency operations involving Taiwan. This fleet is responsible for the area from about 35°-10'N to 23°-30'N, which corresponds to the Nanjing Military Region. In addition to its headquarters at Ningbo, it has bases at Shanghai, Wusong, Dinghai, Hangzhou, Xiamen, and Xiangshan. Its forces include two submarine, two escort, one mine warfare, and one amphibious squadrons. It no doubt would be able to call on the resources of the entire PLAN, in the event of a crisis.[60]

The South China Sea. The SCS is also a maritime issue for Beijing. This relatively small but contiguous oceanic area embodies important economic and political strategic issues. These include rich fisheries; possible large petroleum, natural gas, and manganese reserves; vital SLOCS; and the issue of national pride, which drives Beijing to a rigid policy of strategic primacy in those waters.[61]

Liu Huaqing, when head of the PLAN, noted his service's mission to secure the "vast resources" of the sea.[62] In February 1992, mid-way through his tenure as China's senior uniformed officer, the National People's Congress passed the Law of the Territorial Sea and Contiguous Zones, which included essentially all of the SCS as sovereign Chinese territory, ocean as well as land areas.

The PLAN's South Sea Fleet, headquartered at Zhanjiang, would have to enforce this law. It is responsible for the area from about 23°-30'N to the Vietnamese border, corresponding roughly to the Guangzhou Military Region—and includes the Paracel and Spratly Islands in the South China Sea. The fleet has bases at Shantou, Guangzhou, Haikou, Yulin, Beihai, Huangpu, as well as Zhanjiang—and mans outposts on several SCS islands. Its forces include two submarine, two escort, one mine warfare, and one amphibious squadrons, as well as China's Marine brigade.[63] The SCS has been described as comprising "more than two-thirds of China's territorial waters of more than 3 million square kilometers."[64]

The PLAN has limited resources for projecting sustained power throughout the SCS and China will likely continue its present, dual strategic approach in the area. First, it is pursuing discussions to resolve conflicting claims.[65] Second, however, is Beijing's policy of establishing a presence on disputed reefs and islets by building facilities ranging from navigation markers to structures capable of housing personnel and berthing small ships.[66] This policy of "creeping assertiveness" is certainly working in the case of the Philippines, as China builds military facilities on Mischief Reef and other islets, while continuing diplomatic talks with Manila.[67]

India. Chinese strategic concerns about India include the nuclear threat recently evidenced in the successful testing of nuclear weapons and missiles capable of reaching targets throughout much of China, the Sino-Indian border dispute, and India's location astride the Indian Ocean SLOCs.

China may be establishing a maritime security presence in Burma and nearby waters. Rationale for a PLAN presence in these waters includes two maritime concerns arising from China's dependence on trade and imported petroleum. First, it would position Beijing to influence the vital SLOCs through the Indonesian straits. Second, an unfriendly India, with its large, relatively modern navy, could control the Indian Ocean's SLOCs on which China depends.[68] China's activities in Burma indicate a maritime strategic goal of establishing a presence on the western approaches to the SCS.

Sea Lines of Communication (SLOCs). Continued development of China's economy into the 21st century depends on reliable sources of electrical power—from fossil fuel now and for the foreseeable future. The nation currently imports up to 600,000 barrels of petroleum products a day, an amount certain to grow.[69] This imported petroleum, mostly from the Middle East, arrives over sea lanes that pass through the Indian Ocean, and the South and East China Seas. These routes also pass through several geographic "choke points," including the Luzon and Taiwan Straits, the Strait of Malacca, and the Strait of Hormuz.

China can project almost no naval control over these choke points, except for the Taiwan narrows, but the PLAN may be moving specifically to correct this deficiency, such as the previously noted steps to emplace PLAN presence in the western approaches to the SCS. Beijing believes a strong PLAN is vital to resolving all these (and many other) issues of national security concern.[70]

A Blue-Water Navy?

The PLAN required to carry out Liu Huaqing's strategy would include task groups of missile-firing, power-projection capable ships supported by nuclear-powered submarines and tactical air power,[71] but its strategy would still assume a philosophical base not far from People's War. In other words, Liu and presumably the current generation of senior PLAN strategists visualize a navy that is both technologically advanced and politically dedicated, in a Maoist sense—both "red" and "expert." This will be difficult to accomplish.

Liu's strategy focuses on Taiwan, asserting China's claims to offshore territories and natural resources, defending the homeland against invasion, strategic deterrence, and coast guard-type activities.[72] Strategy, however, is a starting point of national security, not its consummation.

The PLAN of the 1990s has grown in size and capability, but in a haphazard fashion, with its leaders forced by budgetary scarcity, a weak industrial infrastructure, and a lack of time, to pursue various paths to modernity, including building, purchasing, and reverse-engineering platforms and systems. Liu Huaqing has launched a modern maritime strategy, but the PLAN is still far from the modern, power-projecting force needed to carry it out.

Achieving the third stage of Liu Huaqing's strategy—global maritime power—would require China's leaders to adopt as a national goal construction of a very large navy during the next 50 years. This maritime strategic objective, if distant in accomplishment, or even chimerical, may still serve the PLAN in domestic budget battles.

The navy is currently following three paths to modernization: buying, reverse engineering, and indigenously producing new ships, aircraft, and weapons systems.[73] There are several reasons for these different paths. First, domestic military and political concerns limit resources available to the PLAN. Second is what history indicates will be a fairly transitory friendship with Russia.[74] Third is the questionable health of the Chinese economy.[75] Fourth, China's military-industrial infrastructure has far to go to reach 21st century capabilities. Finally, naval planners face China's lack of maritime tradition: voyages half a millennia ago do not constitute a useful heritage, when the intervening centuries have been marked by introspective nationalism.

Building a navy in this incoherent fashion significantly increases the complexity of training, manning, and maintaining the fleet. Just managing an effective parts supply system for a fleet of many small ship classes is difficult, as is outfitting and supporting it logistically. Further, purchasing military equipment abroad is a questionable option in the long term. First, nations rarely sell their most capable front-line weapons to others.

Second, not every nation can produce such weapons, no matter how well-intentioned. For instance, modern tactical jet aircraft can be designed and produced by only very few nations: the United States, Russia, Great Britain, France, Sweden, and perhaps Japan. Modern SSN design and manufacture lie within the capability of even fewer: the

United States, Russia, and perhaps France.[76] Third, buying a modern force takes investment that could otherwise be devoted to building the indigenous infrastructure to do so, which inherently limits a nation's military potential.

PLAN Capabilities. With about 240,000 personnel, the navy is by far the smallest of the PLA services: the army numbers about 1.5 million and the air force approximately 470,000. Yet the PLAN commands perhaps one-third of the military's total budget. The navy has thus fared well within the PLA, but has a large appetite. Modern naval systems are technologically complex and expensive, and require significant human as well as material resources. The army will always be dominant in China's military, but the navy's relatively strong status during current PLA-wide modernization efforts attests to Beijing's concern about maritime issues.[77]

The head of the PLAN, Vice-Admiral Shi Yunsheng, has attributed several features to China's 21st century navy: (1) an "'offshore defense' strategy;" (2) "making the navy strong with science and technology, narrowing the gap between it and other military powers;" (3) "more advanced weapons," including "warships, submarines, fighters, missiles, torpedoes, guns, and electronic equipment;" and, (4) trained personnel and "more qualified people."[78] Notably absent from this list is mention of logistics or sustainment—the ability to keep a fleet at sea for an extended period of time;[79] but these goals illustrate the PLAN leader's appreciation for how far his force has to go to "improve its capacity to win a war at sea."

Doctrine. Defined in the United States as "fundamental principles by which the military forces or elements thereof guide their actions in support of national objectives [and which] is authoritative but requires judgement in application,"[80] doctrine provides the crucial bridge between strategic intent and operational effectiveness. Doctrine presumably is driven by anticipated employment, pictured in "illustrative scenarios" that in turn furnish military planners with schema to frame their planning for future operations.

The United States figures as the *bete noir* in PLAN scenarios, and determination of doctrine, as well as operational planning, must be based on whether the U.S. Navy and Air Force are likely to be involved. Chinese naval strategists appear to understand the vast gulf in capabilities between the PLAN and the U.S. Navy.[81] As a result, PLAN planners should be expected to try to seize the initiative in an operational situation where the United States might be a participant. They will focus on getting in the first blow.

Did the PLAN "learn" lessons from Operation DESERT STORM and again from Kosovo, where U.S. superiority in military technology and operational power was graphically demonstrated? If it did, the chief lesson may well be that Chinese forces will for at least the next decade be incapable of successful direct confrontation with American forces. Instead, the PLAN will have to rely on speed/mobility, flexibility, and preemption in a contest with the United States. These are generalities, however, not **operational** doctrine, which is linked directly to capabilities. Maritime warfare is by nature multidimensional, a characteristic becoming steadily more complex as information-age developments are adapted for naval use. The PLAN is aware of this increasing complexity, but has yet to demonstrate the capability to operate in that environment.

Beijing is trying to create a modern navy capable of carrying out a blue-water maritime strategy, but China's research and development still suffers from 50 years of abrupt, violent political changes that rent the educational and scientific fabric of the country. China currently is capable of building new ships, but ships built in 1999 on a 1969 technological base are obsolete when their hulls first hit the water. They are not necessarily ineffective, but must be employed conservatively and imaginatively.

Acquiring modern ships is relatively easy, if one's checkbook is big enough. But an effective navy requires sophisticated maintenance, supply, and training capabilities that are expensive to develop, far more difficult to learn and, even more importantly, to institutionalize as a naval "way of life." China has recently purchased naval weapons and technology, especially from Russia, that demonstrate Beijing's determination to speed the pace of naval modernization.[82]

Post-GCPR naval expansion has emphasized moving toward a blue-water navy to "respond to the needs of the national strategy and the national defense strategy." The ships and systems emerging from that period are credited with giving "China a significant main naval fighting force" but inadequate "to have all-around (three-dimensional) control of blue water." The PLAN leadership seems to understand what it needs to achieve its strategic goals.[83]

China's ability to project power is key to its maritime strategy and is central to Shi's vision. This capability was not a national security need before the 1985 strategic shift: the Korean, Indian, and Vietnam excursions were overland operations. Projecting power requires the ability to exercise not just sea denial, but the ability to control or significantly affect events ashore from sea-based units. A wide range of activities may qualify, from a demonstration offshore (e.g., aircraft carriers deployed to the vicinity of Taiwan) to launching cruise missiles from submarines against shore targets (e.g., U.S. attacks on Iraq).

The PLAN must also be able to support forces at sea, over time, and requires an amphibious assault capability. China has never possessed a robust capability to transport and land troops under combat conditions, and the PLAN does not appear to be making a dramatic effort to correct this deficiency.[84]

The heart of China's navy is surface combatants, described as the "vital" or "main" PLAN component. The Chinese class their warships "on a par with foreign warships of the 1980s," but that is an optimistic estimate.[85] The PLAN surface forces suffers from ASW, AAW, and ASUW deficiencies, and especially from a lack of effective **area** defense—the ability to defend not just individual ships, but groups of lightly armed transports.

The PLAN has frequently demonstrated its ability to deploy and maintain ships in the South China Sea, although these exercises have not drawn as much attention as those in the Taiwan Strait area. In mid-October 1996, for instance, the PLA reportedly conducted a 15-day exercise "seizing islands."[86] This exercise, as well as those conducted in the vicinity of Taiwan in 1995 and 1996, is clear evidence of the PLAN exercising in support of national strategic objectives. This linkage between training and/or exercising and maritime strategy is more significant than the technological complexity of that training. A 21st century PLAN must possess both the technological sophistication and personnel expertise required to accomplish its strategic goals.[87]

These shortfalls may be at least partially compensated for by innovative operational doctrine. This in turn is driven by the putative revolution in military affairs (RMA), which was demonstrated in Operation DESERT STORM and dramatically emphasized during the U.S. campaign in Kosovo, and appears to offer such an opportunity to some PLA strategists. One has written, for instance, that "cruise missiles are the vanguard, aerial strength is the main power, and the ground, sea, air, space, and electromagnetism are integrated. **This will become a basic mode for the recent and future high-technology regional war**," the battlefield for which "will be a digitized battlefield" (emphasis added).[88]

Nuclear Deterrence at Sea. The PLAN faces financial, industrial, and technological limitations integrating the RMA; nuclear deterrence may be the naval strategic mission to receive priority. China has built two strategic missile submarines: a conventionally powered *Golf*-class (a

1950s Soviet design) and a nuclear-propelled *Xia*-class, neither of which is apparently operational. These ships are both capable of launching the CSS-N-3 1,500 nm-range Intermediate Range Ballistic Missile (IRBM).

China is reportedly building a follow-on class to the *Xia*, the Project-094 boat. This effort may be receiving Russian assistance and has the potential to alter radically the strategic maritime picture in Asia: a Project-094 armed with the new JL-2 ICBM (a possibly MIRVed maritime version of the DF-31 nearing the end of development) would for the first time enable China to put a strategic deterrent to sea that would credibly threaten the United States. The new FBM will likely go to sea during the next decade.[89]

Air Power at Sea. Any effective offshore maritime strategy requires air power, and no aspect of PLAN modernization attracts more interest and generates more concern than the possible acquisition of aircraft carriers. A carrier is seen as providing China with air cover for the long-range power projection needed to seize and hold disputed territory such as the Spratly Islands. And a PLAN carrier force operating east of Taiwan would place that island's air defense forces in the middle of an attack from two fronts, if the PLA were able to coordinate carrier-based attacks with shore-based attacks from the mainland and successfully defend the carrier.

Aircraft carrier acquisition is potentially the PLAN's most important modernization decision. Will the first carrier be a small, 12,000- to 16,000-ton vessel with a "ski jump" bow to facilitate launching its fixed wing aircraft, or will it be a large conventional-takeoff-and-landing (CTOL) 40,000- to 60,000-ton ship with catapults and arresting wires? Building and operating a CTOL carrier would be a very ambitious undertaking that would impact PLA (not just PLAN) force structure in a basic way.

Thailand's purchase of a carrier, coming on the heels of U.S. carriers' arrival in the Taiwan area in March 1996, must have further frustrated PLAN strategists.[90] Never was the effectiveness of aircraft carriers as political instruments more directly felt by Beijing. Indeed, Chinese acquisition of a carrier in the near future, in conjunction with the purchase of anti-carrier *Sovremennys*, will indicate that the United States' 1996 action spurred China's naval modernization on a scale similar to that followed by the Soviet Union after the embarrassment of the 1962 Cuban Missile Crisis. It will also be an indicator that Deng Xiaoping's four national priorities have changed.

In fact, PLAN strategists who favor large carriers may be forgetting that aircraft carriers are a means, not an end. The "end" is correcting the PLAN's most crucial shortcoming—the lack of air power at sea. Solving this deficiency may involve carriers, but it almost certainly does not require large CTOL ships.

Prospects: Beijing's Maritime Strategy.

China faces five major maritime security situations in Asia: Japan, Taiwan, the South China Sea, India, and vital SLOCs. Maritime strategists in Beijing are reminded of their navy's shortcomings every day the U.S. Navy deploys across East Asia—which no doubt helps fuel the current campaign to modernize the PLAN. The navy seeks greater support as a key player within the army-dominated PLA and in the national security policy process. Casting the United States as the adversary facilitates these efforts, given the overwhelmingly maritime nature of U.S. presence in East Asia.

The 14th CCP Congress in October 1992 elected eight admirals to its Central Committee, including Liu Huaqing. Liu retired at the September 1997 15th Party Congress as China's senior uniformed officer, however, which has likely not helped the PLAN's modernization program. Six admirals were named to the Central Committee at this congress. Additionally, the current PLAN commander, Vice-Admiral Shi Yunsheng, is relatively junior.

Although Shi has claimed that the "CPC leadership" believes that "building a powerful People's Navy" is the "major task of our Army building," his ability to prevail in budget contests is not likely to match that of Liu Huaqing.[91] The Chinese Navy remains much smaller than the army, and it is not clear that its disproportionate share of the defense budget has resulted in significantly greater clout within the senior ranks of Chinese strategists.

The PLAN offers China's leaders a flexible, ready instrument of power projection, and Beijing has not hesitated to use the PLAN to achieve strategic goals: witness the 1974, 1988, 1995, and 1998-99 actions in the South China Sea. "Offshore Defense" is a maritime strategy with clear offensive implications. Beijing is moving its strategic line seaward from the coast, demonstrating that the navy has a key role in China's 21st century strategy. Insofar as the PLAN is concerned, a strategy of "offshore defense" includes missions to (1) contain and resist foreign aggression from the seas, (2) defend China's territory and sovereignty, and (3) safeguard the motherland's unification and marine rights.[92]

These strategic objectives translate into complex and difficult specific missions, including: (1) preparing for operations against Taiwan; (2) defending Chinese claims in the East and South China Seas; (3) maintaining a strategic deterrent force against the United States (and possibly India and Russia); (4) protecting vital SLOCS—some lying a great distance from China;[93] and, (5) serving as a diplomatic force.[94]

Beijing's maritime strategy development seeks to include both modern technology and Maoist doctrine, as in "the use of strategy can reverse the balance of combat strength."[95] The ideal maritime strategy will overcome recognized shortcomings in doctrine, equipment, and training.[96] A strategy cannot, however, overcome the shortfalls in mobility and power projection that mark the PLAN.

The PLAN is a long way from being the dominant naval power in East Asia, even apart from the U.S. maritime presence. The JMSDF is certainly superior to the PLAN, and the ROKN would be a very difficult opponent. Even the Taiwan navy would not be a pushover for the PLAN. Clearly, a wise maritime strategist in Beijing would not, in the event of conflict, position the PLAN "one-on-one" against any of these modern naval forces.

A more thoughtful strategy would be required for the PLAN to achieve specific goals in the face of opposition by the USN, the JMSDF, the ROKN, or the Taiwan Navy. One aspect of such a strategy would almost certainly be employment of information warfare methods to counter the advanced military technological superiority of these fleets—a capability more discussed than demonstrated in China.

Another strategic step in such a conflict would be to gain the initiative through preemption. This does not necessarily require a "bolt from the blue,"[97] but could be achieved by seizing the initiative at a time of significant naval weakness on the part of the adversary.

The PLAN's strategic responsibilities are challenging. It does not have the resources to defend distant SLOCs, and the distances involved in securing the South China Sea are daunting to a navy weak in air power, amphibious lift, and logistics sustainment. The opposition posed by Taiwan makes any assault on that island a significant military as well as political problem. And in the Japanese navy, the PLAN would face a superior adversary. To be effective, China's maritime strategy must compensate for the PLAN's material shortcomings and its lack of operational experience. That said, the PLAN today is a formidable force within littoral East Asia, and is viewed as a vehicle for aggression.[98]

Some analysts combine thoughts of the RMA with China's extensive maritime claims to conclude that China needs to establish the maritime power necessary to defend its interests, including Taiwan and the SCS.[99] One such view from Beijing describes the sea as the "New High Ground of Strategic Competition" and focuses PLAN attention on five areas of international rivalry, those over ocean islands, sea-space jurisdiction, marine resources, maritime strategic advantage, and strategic sea-lanes. The seas are described as both "a protective screen" but also as "a marine invasion route." Naval missions are seen as first, coastal defense; second, control of the "sea space," which is "four dimensional," including air, surface, subsurface, and the seabed. The Asia-Pacific region is defined as a "priority region of maritime strategic competition."

> Control of the seas involves many areas, including political, economic, diplomatic, science and technology, and military. Military control of the seas means achieving and defending national unification, defending national maritime territorial sovereignty and maritime rights and interests, protecting legitimate maritime economic activities and scientific research, and ensuring a peaceful and stable climate for national reform, opening, and coastal economic development, by dealing with possible maritime incidents, armed conflicts, and local wars.... our Navy has an inescapable mission.... The 21st century is going to be a maritime one.... we will have to make our maritime strategy a key part.[100]

Today and for the next decade, given the presence of peaceful borders to the north and west, Beijing's national security priorities lie to the maritime east and southeast. The PLAN must be able to control East Asian seas to facilitate accomplishment of Beijing's strategic aims, but China faces significant hurdles in the technological and industrial infrastructure and resource availability needed to deploy a regionally dominant navy.

A strategist in Beijing faces several strategic priorities.

First may be a national imperative to establish an effective nuclear deterrent force at sea as the core of a 21st century maritime strategy. Second will be maintenance of a naval presence throughout East Asia, using port visits to the nations of Northeast and Southeast Asia, with an occasional foray to Southwest Asia and the Western Hemisphere. Within this general policy of presence, the PLAN will be focused, as part of a joint force with the PLAAF, on specific objectives, starting with Taiwan. This in turn requires a credible power-projection force, with enough amphibious and logistics capability to take and hold disputed territory in the East and South China Seas. Fifth, SLOC defense will be pursued; likely avenues are expanding a presence west of Malacca and diplomatic efforts in Southwest Asia.

Finally, the question of Taiwan will dominate national security strategy discussions. PLAN strategists will have to ensure that their force remains a key player in plans to coerce Taiwan's reunification, but do so in a manner that will not consume all modernization efforts. Given present and probable future budget resources, it would not be in the PLAN's interest, for example, to have to build a massive amphibious force.

The first stage of Liu Huaqing's reported strategy—to control China's adjacent seas out to the "first island chain," is reasonable and attainable within the next 20 years, but only if Beijing changes the national prioritization of resource allocation necessary to build a modern maritime force. There is currently insufficient evidence—acquisition of technology and a few units of disparate ship classes do not qualify—to conclude that such a significant shift is being made.[101] Comparison to Germany's late 19th century naval building "dash" to catch up to Great Britain's navy is not a valid analogy of what China could do. Imperial Germany already possessed in the late 1890s an industrial-technological-scientific infrastructure equal and perhaps superior to Britain's: China fails completely to match this status. Also, Germany built the wrong navy; it was unable to serve as more than a coastal defense force and, in the final analysis, served little more purpose than to absorb vast quantities of national resources while needlessly alienating Great Britain and other powers.[102]

Conclusion.

Earlier, I posed a list of nine factors that affect a nation's development of a maritime strategy. How does China measure up? First, training and education programs are receiving increasing attention as the professional specialization of the officer corps increases. Progress undeniably is being made to adopt modern training and education systems and methodology, but suffers from time devoted to political education and resource limitations.

Second, naval systems and platform costs, capabilities, and sustainability are the focus of PLAN modernization efforts. As new systems and platforms are bought on the global market and produced in China, the PLAN suffers from resource limitations, a relatively weak indigenous infrastructure, and low starting point for modernization.

Third, the national scientific and industrial infrastructure for research, development, and production of naval warfare technology and systems is improving, but remains inadequate to support rapid design and buildup of state-of-the-art systems from drawing board to operational force in a timely manner. Hence the continued reliance on foreign purchases.

Fourth, the ability to derive doctrine and tactics is uncertain, but clearly advancing, as evidenced in publications, military education and training, and exercises, especially those focused on joint operations, use of information warfare (IW), and integrated systems employment.

Fifth, the ability to administer, operate, and command and control tactical units beyond individual ships is improving, as demonstrated by current administrative streamlining and recent long-range deployments of small flotillas, but is still a question mark on a fleet or theater level. Sixth, sources of intelligence, and its production, analysis, and dissemination absorb major resources and may be a mainstay of PLAN modernization.

Seventh, service-wide naval strategic planning appears to be ongoing, with apparent focus on trying to avoid rather than on matching a potential adversary's (i.e., the United States) strengths. Eighth, naval influence at the national level is probably weaker following Lui Huaqing's retirement. The awareness of maritime power is likely being increased, however, based on China's increasing dependence on international trade and commercial shipping.

Finally, there seems no evidence that naval strategists hold an enhanced position in the national strategy-making structure. Their status is likely to rise, however, in proportion to the degree of crisis in maritime situations, such as Taiwan or the South China Sea.

China is pursuing a conscious maritime strategy designed to achieve near-term national security objectives and longer-term regional maritime dominance through both combatant and merchant fleets. In the near term, Beijing is building a navy capable of decisively influencing the operational aspects of the Taiwan and South China Sea situations, should diplomacy and other instruments of statecraft fail. Establishing PLAN or PLAN-AF bases in the Spratlys (perhaps by artificially enlarging one of the land forms) and on Burmese territory would provide a starting point for a major Chinese role in controlling the Malacca and associated straits, key to the economic life of East Asia.

Power-projection ashore, with its emphasis on littoral warfare, also tends to lessen the traditional importance placed on "blue-water operations," ironically at the very time when China seems focused on developing a navy specifically for that arena.[103] Furthermore, in addition to the problem of pursuing multiple paths to modernization, the PLAN is caught up in a web of ideology and domestic politics that complicates the inherently expensive and complex process of building a navy. The shorthand terms "red versus expert" or "ideology versus technology" serve to sum up the problem faced by Chinese strategists trying to build a large, modern navy—be it in the 15th, 19th, or late 20th centuries. It is a situation to be welcomed by China's maritime neighbors.

CHAPTER 8 - ENDNOTES

1. Quoted in Cha Chun-ming, "Chinese Navy Heads Toward Modernization," *Ta Kung Pao*, Hong Kong, April 11, 1999, p. B6, in *Foreign Broadcast Information Service-China* (henceforth *FBIS-CHI*)-1999-0418.

2. John W. Garver, *Foreign Relations of the People's Republic of China*, Englewood Cliffs, NJ: Prentice-Hall, 1993, pp. 250-252.

3. Nan Li, "The PLA's Evolving Warfighting Doctrine, Strategy and Tactics, 1985-95: A Chinese Perspective," *China Quarterly*, No. 146, June 1997, p. 402, notes that "Since the mid-1980s, China has claimed sovereignty over three million square kilometers of maritime territory (China's land territory is 9.6 million square kilometers), . . . This territory covers 320 kilometers of continental shelves and exclusive economic zones, and extends 1,600 kilometers to include the whole of the Nansha (Spratly) Islands."

4. Maritime distances throughout this chapter are given in nautical miles, one of which equals approximately 1.2 statute miles.

5. Although, as pointed out by RADM J.R. Hill, USN (Ret.), *Maritime Strategy for Medium Powers*, Annapolis, MD: U.S. Naval Institute Press, 1986, p. 85, if China were to employ this strategy, say in the case of Taiwan, "the penalties for getting it wrong may be quite severe." (p. 85). Hill also delineates five indicators of "sea dependence," all of which apply to China: seagoing trade, fish catches, size of Merchant Marine, ship building and repairing, and the offshore zone. (p. 229).

6. One can buy the ships and systems that make up a navy, but they are almost worthless without the infrastructure to maintain and support them, as well as the operational know how to employ them effectively.

7. Cited in David Alan Rosenberg, "Process: The Realities of Formulating Modern Naval Strategy," conference paper, The Corbett-Richmond Conference, Naval War College, Newport, RI, September 28-29, 1992, p. 3.

8. This list is a variation on the 17 points delineated in *Ibid.*, pp. 6-20.

9. Colin S. Gray, "Seapower and Landpower," in Colin S. Gray and Roger W. Barnett, eds., *Seapower and Landpower*, Annapolis, MD: U.S. Naval Institute, 1989, p. 4.

10. *Ibid.*, p. 13.

11. Jiang Zemin, Speech at Geneva, March 26, 1999, quoted in *Xinhua*, Beijing, March 26, 1999, in *FBIS-CHI*-99-0326. Jiang is also quoted in the *Xinhua Domestic Service,* Beijing, April 26, 1999, in *FBIS-CHI*-99-0406, as describing "the current international situation [as] relaxing, but . . . The world is not peaceful."Also see Luo Renshi, "What is New About the 'New Gunboat Policy'," *Jiefangjun Bao*, Beijing, May 20, 1999, p. 5, in *FBIS-CHI*-99-0526, who attributes this American policy to a "Gulf War Syndrome" replacing the "Vietnam War Syndrome."

12. China is a net energy (and a net petroleum) importer; its ability to feed its still growing population is hampered by the low proportion of its land (about 10 percent) that is arable.

13. Dong Guozheng, "Security Globalization is Not Tantamount to Americanization," *Jiefangjun Bao,* Beijing, May 24, 1999, p. 5, in *FBIS-CHI*-99-0602.

14. Theater missile defense (to include coverage of Taiwan) is described as another U.S. instrument for "actively pushing its 'neo-interventionism' to dominate the world," through a "two-flank encirclement of Russia and China." The United States is accused of wanting to control Taiwan since it is China's "most direct door . . . to the Pacific," and the Spratlys, since they offer China "a strategic base closest to the Strait of Malacca." A variation on this theme is the idea that the United States is "hatching Six Major Conspiracies Against China" by "creating trouble and stirring up unrest in China," using Japan "to create disturbances in Asia [and] to intervene in events occurring in areas surrounding China," "attempting to drag China into the mire of the arms race" through pursuing TMD, "preparing to intervene in the war in the Taiwan Strait," and "trying to undermine China's national policy of reform and opening up."

The thread of continuity throughout these views is the United States using TMD and a militaristic Japan to help enforce the Asian part of a policy of global hegemony, in part by containing China. See, for instance, Li Donhang, "Dangerous Attempt to Resist Multipolarization

Process," *Jiefangjun Bao,* May 26, 1999, p. 5, in *FBIS-CHI*-99-0604; Li Li, "Monster of Dark Civilization and Conspiracy Strategy—Commenting on Cox Report," *Jiefangjun Bao,* Beijing, May 29, 1999, p. 4, in *FBIS-CHI*-99-0604; and Zhang Dezhen, "On U.S. Eurasian Strategy," *Renmin Ribao,* Beijing, June 4, 1999, p. 6, in *FBIS-CHI*-99-0605.

15. See Michael Krepon, ed., *Chinese Perspectives on Confidence-building Measures*, Washington, DC: The Henry L. Stimson Center, 1997, for the sophistic arguments by four Chinese authors that China should not be held to international norms of "transparency" in national security matters, since the country is still a "developing nation."

16. Jen Hui-wen, "Jiang Zemin Talks Again on Strengthening the Army Through Science and Technology," *Hsin Pao,* Hong Kong, April 16, 1999, p. 16, *FBIS-CHI*-99-0417.

17. Li Dan, report of Zhang Wannian speech in *Cheng Ming,* Hong Kong, April 1, 1999, pp. 21-22, in *FBIS-CHI*-99-0414. *FBIS* characterizes this source as a magazine with a "tendency to sensationalize."

18. David M. Finkelstein, "China's New Security Concept: Reading Between the Lines," Center for Naval Analysis (CNA) Issue Paper, Alexandria, VA: CNA, April 1999, provides an interesting discussion of why Beijing is unhappy with the present international environment.

19. Paul H.B. Godwin, "Force Projection and China's Military Strategy," conference paper, Sixth Annual Conference on the PLA, Coolfont, Virginia, June 1995, p. 4. Also see Godwin, "Changing Concepts of Doctrine, Strategy, and Operations in the People's Liberation Army 1978-87," *China Quarterly*, No. 112, December 1987, pp. 573-590.

20. The Ming decision also reflected Chinese xenophobia, perhaps best expressed in the Qing Emperor Ch'ien-lung's response to Britain's 1793 attempt to establish relations with Beijing when he told Lord McCartney that "we possess all things. I set no value on objects strange or ingenious, and have no use for your country's manufactures."

For an interesting but Eurocentric interpretation of the role maritime mobility played in European imperialism, see George Raudzens, "Military Revolution or Maritime Evolution? Military Superiorities or Transportation Advantages as Main Causes of European Colonial Conquests to 1788," *The Journal of Military History*, Vol. 63, No. 3, July 1999, pp. 631-642.

21. PLAN Vice-Commander Zhou Xihan, 1957, quoted in David G. Muller, Jr., *China as a Maritime Power,* Boulder, CO: Westview Press, 1983, p. 47.

22. Shu Guang Zhang, *Mao's Military Romanticism: China and the Korean War, 1950-1953*, Lawrence, KS: University Press of Kansas, 1995, pp. 46-54, provides a useful description of the beginnings of the PLAN.

23. About 2,000 former Republic of China naval personnel who defected to the communist regime in 1949 formed the core of the nascent PLAN. See Muller, p. 13.

24. Vladimir Lenin, cited in Bruce W. Watson, "The Evolution of Soviet Naval Strategy," in Bruce W. Watson and Peter M. Dunn, eds., *The Future of the Soviet Navy: An Assessment to the Year 2000*, Boulder, CO: Westview Press, 1986, p. 115.

25. *Ibid.*, p. 15.

26. Raymond V.B. Blackman, ed., *Jane's Fighting Ships: 1955-56*, London: Jane's Fighting Ships Publishing Co., 1956, pp. 151ff provides these numbers, but they should be treated only as estimates.

27. Bruce Swanson, *Eighth Voyage of the Dragon: A History of China's Quest for Seapower*, Annapolis, MD: Naval Institute Press, 1982, p. 196.

28. Jun Zhan, "China Goes to the Blue Waters: The Navy, Seapower Mentality and the South China Sea," *The Journal of Strategic Studies*, Vol. 17, No. 3, September 1994, p. 187.

29. Muller, p. 16.

30. See Robert J. Donovan, *Tumultuous Years: The Presidency of Harry S. Truman, 1949-1953*, New York: W.W. Norton, 1982, p. 206, for Truman's decision to reposition the Seventh Fleet, and pp. 241ff for a good account of administration thinking (Truman, Acheson, Bohlen, *et al.*) about the implementation of NSC-68, which effectively re-armed the United States for the Cold War and potential global war with Soviet-led communist forces: "On the last day of July 1950, Truman and Acheson had a talk about grand strategy. The eyes of the American people were glued to Korea. . . . The president and the secretary of state fixed their gaze on the Rhine and the Elbe." (Donovan, p. 244.)

31. Swanson, p. 206.

32. See John Wilson Lewis and Xue Litai, *China's Strategic Seapower*, Stanford: Stanford University Press, 1984, pp. 206ff.

33. Blackman, ed., *Jane's Fighting Ships: 1960-61*, New York: McGraw-Hill, 1961, pp. 117ff.

34. This brief discussion of the early development of maritime thought in the People's Republic of China relies on Swanson, pp. 187-236.

35. China built two *Xia*-class fleet ballistic missile submarines, patterned on the U.S. *George Washington*-class/Soviet *Hotel*-class. One of the *Xia*s apparently was so severely damaged in a fire that it was not completed; the second successfully launched a missile in 1985, but may have never made any patrols at sea. (See Richard Sharpe, ed., *Jane's Fighting Ships: 1995-96*, London: Butler and Tanner, 1996, p. 114.)

36. Presumably, the United States would have come to Taiwan's defense had the PRC tried to take advantage of the American preoccupation with Vietnam by attacking the island, but the Great Proletarian Cultural Revolution (GPCR) was even more of a preoccupation for Beijing.

37. *FBIS* reports, cited in Muller, p. 154.

38. Lewis and Xue, pp. 147-148, 223.

39. The PLAN also included more than 30 other subs, a collection of assorted foreign-built destroyers and escort vessels (Soviet, Japanese, U.S., British, Canadian, and Italian), and more than 400 Chinese-built patrol craft, some of them hydrofoils and most armed with torpedoes. See Blackman, ed., *Jane's Fighting Ships: 1970-71*, New York: McGraw-Hill, 1971, pp. 61ff.

40. Quoted in Kenneth R. McGruther, *The Evolving Soviet Navy*, Newport, RI: Naval War College Press, 1978, pp. 47-48, 66-67.

41. Muller, p. 171.

42. John E. Moore, ed., *Jane's Fighting Ships: 1976-77*, New York: Franklin Watts, 1977, pp. 100ff. The PLAN also included the first Chinese range instrument ships, for tracking guided missile flights, and the first Chinese-built amphibious transports.

43. As CMC Vice-Chairman, however, Liu maintained a balanced view of PLA needs, apparently sacrificing a CV building plan to ensure

that research and development funds were spread evenly among the services. Alfred Wilhelm, conversation with author, September 1999.

44. See Alexander C. Huang, "Chinese Maritime Modernization and Its Security Implications: The Deng Xiaoping Era and Beyond," unpublished Ph.D. dissertation, The George Washington University, Washington, DC: 1994, pp. 225ff for this discussion. The dates given for achieving Liu's three stages vary from source to source; no authoritative source has been identified, as pointed out by Taeho Kim in discussion with the author, September 1999.

45. See John Downing, "China's Evolving Maritime Strategy," *Jane's Intelligence Review,* March 1, 1996, p. 130 and Huang, dissertation, p. 230.

46. Cited in Chen and Chai, *FBIS-CHI*-97-329. Also see Alexander Huang, "The Chinese Navy's Offshore Active Defense Strategy: Conceptualization and Implications," *Naval War College Review*, Vol. 47, No. 3, Summer 1994, pp. 16ff, for a good discussion of the "First" and "Second Island Chains."

47. RADM J.C. Wylie, USN (Ret.), *Military Strategy,* 1967, reprint, Westport, CT: Greenwood Press, 1980, p. 49, is a classic work on modern naval strategy. Also see James A. Winnefeld, "Why Sailors are Different," U.S. Naval Institute *Proceedings*, Annapolis, MD: May 1995, p. 66: "The soldier shapes and exploits his environment; the sailor must adjust to it."

48. See Paul H.B. Godwin, "Technology, Strategy, and Operations: The PLA's Continuing Dilemma," article draft, for a discussion of Chinese efforts to improve warship system capabilities, and Godwin, "From Continent to Periphery: PLA Doctrine, Strategy and Capabilities Towards 2000," *China Quarterly*, No. 146, June 1996, pp. 464-487.

49. Godwin, "Technology, Strategy, and Operations," p. 191.

50. Three ships are normally required to maintain one on patrol at all times, although an FBM armed with a missile capable of hitting its target without going very far to sea (e.g., the Russian *Typhoon*) might not need this two ship "backup."

51. Quoted in Watson, p. 120.

52. The Soviet influence is also reflected in the PLAN's training paradigm, which emphasizes inport training, whereas the U.S. and other western (and Japanese) navies believe that training at sea is necessary for combat proficiency.

53. Cited in Huang, dissertation, p. 99.

54. Quoted in You Ji, "A Test Case for China's Defence and Foreign Policies," *Contemporary Southeast Asia,* Vol. 16, No. 4, March 1995, p. 379.

55. Liu Huaqing has used this argument, stating that "the Chinese navy must live up to the historical responsibility to grow rapidly up into a major power in the Pacific area in order to secure the smooth progress of China's economic modernization." Quoted in Jun Zhan, "China Goes to the Blue Waters: The Navy, Seapower Mentality and the South China Sea," *The Journal of Strategic Studies,* Vol. 17, No. 3, September 1994, p. 191.

56. See *Ping Kuo Jih Pao,* Hong Kong, March 7, 1997, p. A20, in *FBIS-CHI*-1997-114, for the State Oceanography Bureau's estimate of China's maritime wealth as "three million square kilometers of sea areas rich in fishing, petroleum, and mineral resources." Fujian Province reported in 1998 that "more than 900 deep-sea trawlers had been dispatched [around the world], bringing home a total catch of 600,000 tons" of fish—which was considered unsatisfactory. *Xinhua,* Beijing, June 23, 1999, in *FBIS-CHI*-99-0623.

57. See the very informative work by Navy Lieutenant Commander Wayne R. Hugar, "The Sea Dragon Network: Implications of the International Expansion of China's Maritime Shipping Industry," unpublished thesis, U.S. Naval Postgraduate School, Monterey, CA, 1998, pp. xvi, 70.

58. The International Institute of Strategic Studies (IISS), *The Military Balance, 1998/99,* London: Oxford University Press, 1999, p. 180.

59. Zhu Rongji reiterated this during his visit to Washington, D.C. in April. See "Stratfor's Global Intelligence Update," *http://alert @stratfor.com*, April 14, 1999, p. 2.

60. *The Military Balance,* p. 180.

61. The most optimistic estimate of SCS petroleum reserves, 55 billion tons, is in *Xinhua,* Beijing, September 5, 1994, in *FBIS-CHI*-1994-172.

62. South China Sea incidents are discussed, and Liu is quoted in John W. Garver, "China's Push Through the South China Sea: The Interaction of Bureaucratic and National Interest," *China Quarterly,* No. 132, December 1992, pp. 1022-1023. Also see the comprehensive

discussion of the South China Sea situation in Mark Valencia, *China and the South China Sea Disputes*, Adelphi Paper 228, London: IISS, 1995.

63. *The Military Balance*, p. 180. A second marine brigade is reportedly being organized. Author's conversations with PLA officers and American analysts.

64. Han Haibing and Li Xiangdong, "The 50th Anniversary of a Heroic Fleet," *Zhongguo Xinwen She,* Beijing, April 23, 1999, in *FBIS-CHI*-99-0428.

65. See *Xinhua,* Beijing, July 23, 1996, in *FBIS-CHI*-1996-144, for Foreign Secretary Qian Qichen's conciliatory statement at the ASEAN Regional Forum that the situation in the Spratlys was "stable."

66. Mischief Reef (*Huanyidao* in Chinese), located well within the Philippines' 200-nm economic zone (EEZ), is the prime example of this encroachment. Whether the facilities will survive the first significant typhoon to blow through is another question. The United Nations Convention on the Law of the Sea (UNCLOS), Art. 121, specifies that to be considered to have an "exclusive economic zone or continental shelf," land must be able to "sustain human habitation or economic life of their own." Hence, the efforts by various claimants to SCS rocks and reefs to construct facilities capable of "sustaining human habitation."

67. A good discussion of the China-Philippines imbroglio is provided in Ian James Storey, "Creeping Assertiveness: China, the Philippines and the South China Sea Dispute," *Contemporary Southeast Asia*, Vol. 21, No. 1, April 1999, pp. 95ff. The best overall account of the South China Sea disputes remains Valencia.

68. Beijing is following at least three tracks regarding its strategic maritime relationship with Rangoon. First are increased sales of military equipment, accompanied by military technical and operational guidance and advisors. This has ranged from small arms to ocean-going warships, including ten *Hainan*-class coastal patrol craft, with another ten of these boats and two much more capable *Jianghu*-class frigates in the offing (Micool Brooke, "The Armed Forces of Myanmar," *Asia Defence Journal,* January, 1998, p. 14). Second, China has established a military presence in Burma, reportedly building/improving naval support facilities at Hainggyi, Akyab, and in the Mergui Islands off Burma's isthmian coast. Rahul Ray-Choudhury, "Trends in Naval Power in South Asia and the Indian Ocean During the Past Year," *SAPRA India Monthly Bulletin,* January 1996, available at *http://www.subcontinent.com/sapra/96jan/si019603*. Third, China

has been active in an electronic monitoring facility in the Coco and Hangyi Islands in the Andaman Sea that may be able to provide Beijing's naval commanders with valuable intelligence on Indian and other naval forces in the Indian Ocean. See Barbara Opall-Rome, "Chinese Moves Roil Region," *Defense News*, February 8, 1999, p. 1.

69. Daniel Yergin, Dennis Eklof, and Jefferson Edwards, "Fueling Asia's Recovery," *Foreign Affairs*, Vol. 77, No. 2, March/April 1998, p. 42, estimate China will import "as much as 3 million" barrels of oil per day by 2010. This article also provides a useful survey of possible alternative Chinese energy sources, including oil fields in the Tarim Basin and Kazakstan, and investment in Venezuelan, Iraqi, Iranian, and Indonesian fields. Also see Hugar, pp. 22ff, for a discussion of Chinese oil imports.

70. See, for instance, Si Yanwne and Chen Wanjun, "Navy to Develop More High-Tech Equipment, *Jiefangjun Bao,* Beijing, June 9, 1999, in *FBIS-CHI*-99-0611, quoting General Cao Gangchuan, director of the General Armaments Department that "it is necessary to put [Navy] armament development in a prominent position of army building . . . . increase armaments' scientific and technological contents; and improve the quality and speed of armament development," and *Xinhua*, Beijing, June 10, 1999, in *FBIS-CHI*-99-0609, quoting Cao that "The navy's rapid reaction capacity, emergency field repair ability and defense readiness must also be improved."

71. See John Wilson Lewis and Xue Litai, "China's Search for a Modern Air Force," *International Security*, Vol. 24, No. 1, Summer 1999, p. 10, for Deng Xiaoping's statement that "The army and the navy both need air cover . . . . Without air cover, winning a naval battle is also out of the question."

72. *Xinhua*, Beijing, June 18, 1999, in *FBIS-CHI*-99-0618, reported that Shanghai had established a "Maritime Safety Administration, the first of its kind in China's coastal areas, . . . to supervise the management of navigation marks, the surveying of sea-routes, and the inspection of ships and maritime facilities."

73. Christopher D. Yung, "People's War at Sea: Chinese Naval Power in the Twenty-First Century," Center for Naval Analysis, Alexandria, VA: 1995, contains detailed discussion of Chinese options for acquiring a large modern navy through either overseas purchase, reverse engineering, or indigenous development.

74. Sino-Russian relations since the 13th century have been mostly adversarial. Even the post-1918 alliance lasted just half a decade, that

after 1949 barely a decade, and the present "strategic" relationship exists largely because of mutual resentment of the United States, rather than as a result of shared values or extensive mutual material or ideological interests.

75. See Nicholas R. Lardy, *China's Unfinished Economic Revolution,* Washington, DC: The Brookings Institution, 1998, for the best current analysis of China's economy. Lardy notes the high percentage of non-performing loans, social and political costs of putative reforms, the weak financial position of state-owned enterprises, and a marked decline in government revenues as evidence of the Chinese economy's questionable health.

76. See Bernard D. Cole and Paul H. B. Godwin, "The PLA and Technology in the 21st Century," conference paper, Wye River Conference on the PLA, September 1998, for a discussion of China's current capabilities in technology applicable to military systems development. The point here is not so much the existence of the technology, as it is the ability of China's military-industrial infrastructure to develop operational military systems from that technology.

77. A recent U.S. Defense Attaché in Beijing has stated that the PLAN and PLAAF have been receiving most of the recent PLA modernization funding. Lewis and Xue, "China's Search for a Modern Air Force," p. 11, claim that the 1985 strategic shift to non-nuclear war scenarios gave the air force and the navy "pride of place" within the PLA. Also see Huang, "The Chinese Navy's Offshore Active Defense Strategy," p. 9 (Table 1), for the estimate that 32.7 percent of the PLA's 1993 budget was allocated to the PLAN.

78. Quoted in *Xinhua,* Beijing, April 21, 1999, in *FBIS-CHI*-99-0421.

79. Geoffrey Till, "Maritime Strategy in the Twenty-First Century," in Geoffrey Till, ed., *Seapower: Theory and Practice,* Portland, OR: Frank Cass, 1994, p. 193, points out that the United States was able to maintain a task force off the coast of West Africa for **7 months** in 1990-91, before finally evacuating civilians from strife-torn Liberia.

80. *The Joint Staff Officer's Guide,* Armed Forces Staff College Publication No. 1., Washington, DC: National Defense University Press, 1997, p. 0-16.

81. Finkelstein, p. 3, makes this point. Two examples are submarine-launched ICBMs and fleet air defense. In the first case, China may have once, in 1985, successfully launched a missile from its

nominally operational FBM submarine; the United States in March 1999 conducted its 81st consecutive successful launch and flight to target of an ICBM, in that case a Trident D-5. In the second case, while the PLAN's newest warship, the *Luhai*-class destroyer, and its prospective *Sovremenny*-class ships purchased from Russia are equipped with only a limited-range air-defense missile system, the United States is already planning the follow-on to *Aegis*, even now far more capable than any other AAW system in the world. See U.S. Navy Staff, *Undersea Warfare,* Washington, DC: U.S. Navy, 1999, p. 13: two salvos of two missiles each were launched simultaneously in this latest test. Also see Robert Holzer, "Study: U.S. Navy Must Go Beyond *Aegis* Radar Era," *Defense News,* June 7, 1999, pp. 3, 36.

82. Most notable are the *Kilo*-class submarines and *Sovremenny*-class destroyers armed with the potent SS-N-22 cruise missiles, but also important is the "softwear," including the services of Russian engineers, that has been acquired by the Chinese.

83. Vice-Admiral Shi Yunsheng, cited in Huang, Chen, and Zhang, "China Enhances the Navy's Comprehensive Strength—Interview with Naval Commander vice Admiral Shi Yunsheng," *Liaowang*, Beijing, April 1999, pp. 13-15, in *FBIS-CHI*-99-0513. Also see Zhang Wei, *Jianchuan Zhishi*, Beijing, January 1997, pp. 8-9, in *FBIS-CST*-97-006.

84. It appears that the defense budget allocation process is no friendlier to the "Gators" of the PLAN than it is in any other country's navy. Troop-carrying ships, no matter how vital to the power projection mission, simply have neither the glamour nor the profit margin of nuclear-powered submarines or guided-missile destroyers. This may be one reason why the PLAN is only slowly strengthening its amphibious warfare capability.

85. Chuan, *FBIS-CHI*-99-024.

86. *Jiefangjun Bao*, Beijing, October 24, 1996, p. 1, in *FBIS-CHI*-1996-210, reported that this exercise included joint operations among "the three armed services and the issues of coordination and supplies," and "involved sea-crossing, changing ships, and ship formations," with the troops fighting "an integrated war and [attacking] the enemy from the three dimensions." Also see Zhang Zenan's interview with Senior Captain Lin Shuangqiao, "Developing and Expanding Chinese Landing Ship Forces," *Jianchuan Zhishi (Naval & Merchant Ships)*, Beijing, January 6, 1997, pp. 10-11, in *FBIS-CHI*-1997-051: Lin is identified as the "army's chief of staff for naval operation" and an amphibious warfare expert. He describes the "vertical assault" aspect of amphibious landings and claims that the

PLA demonstrated this capability in October 1995 and March 1996 exercises near Taiwan.

87. VADM Shi Yunsheng, listed five attributes of a modern navy: (1) strengthened "research on naval strategies," (2) "vigorous development of high-tech equipment," (3) train PLAN personnel "with modern and scientific and technological qualities" to operate its "modern equipment," (4) effective "medium and long-term" plans, and (5) "modernization of the main equipment of the navy," all of which pose "a new challenge" for the PLAN. Reported in Chen Wanjun and Zhang Chunting, "Shouldering the Important Task of a Century-Straddling Voyage—Interviewing Newly Appointed Navy Commander Lieutenant General Shi Yunsheng," *Liaowang*, Beijing, February 24, 1997, pp. 8-9, in *FBIS-CHI*-1997.

88. Wang Zudian (identified as a Space Technology Information Research Institute researcher), "The Offensive and Defensive of High-Technology Arms Equipment," in *Liaowang Weekly*, 21st ed., quoted in *Xinhua*, Hong Kong, May 24, 1999, in *FBIS-CHI*-99-0526.

89. U.S. Navy, *Worldwide Submarine Challenges, 1996*, Washington, DC: Office of Naval Intelligence, February 1996, p. 27. Predicting the year in which the next PLAN FBM will actually go to sea on deterrent patrol is a risky proposition, given China's poor track record in this area. Two PLA lieutenant-generals, conversations with author, May 1996.

90. The flexibility of aircraft carriers is emphasized in this case since the first to be sent to Taiwan, the USS *Independence*, is homeported in Yokosuka, Japan and was already at sea. The second, USS *Nimitz*, was en route to her West Coast homeport from a deployment to the Persian Gulf and deviated from an intended port call in western Australia to steer for Taiwan. See "Jet Deal Should not Just Fly Through," *Bangkok Post*, July 2, 1999, in *FBIS-EAS*-1999-0702, for a report that due to its reduced budget resulting from Thailand's economic malaise, the navy has been unable to maintain *Chakri Naruebet's* aircraft and cannot afford the fuel to send it to sea.

91. Chen Wanjun and Zhang Chunting, "Shouldering the Important Task of a Century-Straddling Voyage—Interviewing Newly Appointed Navy Commander Lieutenant-General Shi Yunsheng," *Liaowang*, Beijing, No. 8, February 24, 1997, p. 2, in *FBIS-CHI*-1997.

92. Chen and Zhang, p. 3. Shi also calls for a "scientifically feasible . . . Navy development strategy" as part of the "defense development strategy . . . subject to the national development strategy" (p. 5).

93. China ratified the UNCLOS in 1996, and some strategists use this pact as rationale for including "military control of the seas [as] legitimate maritime economic activities." Li Jie and Xu Shiming, "The UN Law of the Sea Treaty and the New Naval Mission," *Hsien-Tai Chun-Shih,* Beijing, February 1997, quoted in Hugar, p. 73.

94. See Lu Ning, *The Dynamics of Foreign-Policy Decisionmaking in China,* Boulder, CO: Westview Press, 1997, pp. 126ff, for an interesting description of the 1988 naval conflict with Vietnam when, according to the author, PLAN forces exceeded and drove national strategy.

95. The question of "red vs. expert" is a facet of civil-military relations, a topic we only imperfectly understand; drawing too sharp a dichotomy between army loyalty (to state, government, or party) and professionalism should be avoided. See works on this topic by David Shambaugh, Michael Swaine, Harlan Jencks, Ellis Joffre, and Jeremy Paltiel.

96. Senior Colonel Huang Xing and Senior Colonel Zuo Quandian, "Holding the Initiative in Our Hands in Conducting Operations, Giving Full Play to Our Own Advantages To Defeat Our Enemy—A Study of the Core Idea of the Operational Doctrine of the PLA," *Zhaongquo Junshi Kexue (China Military Science),* No. 4, November 20, 1996, pp. 49-56, in *FBIS-CHI*-1997. The authors, who both serve at the Academy of Military Science, clearly identify the United States as "our enemy" (p. 8), but display an imperfect knowledge of American weapons systems.

97. "Bolt from the blue" was a concept that first appeared in modern maritime strategy in the early 20th century to describe a possible surprise German naval attack on Great Britain. See Arthur M. Marder, *The Road to War, 1904-1914,* vol. 1 of *From the Dreadnought to Scapa Flow: The Royal Navy in the Fisher Era,* London: Oxford University Press, 1961, p. 144, and S.W. Roskill, *The Strategy of Seapower: Its Development and Application,* Westport, CT: Greenwood Press, 1962, p. 104, for discussion of this concept.

98. There is little question that this is the view from Manila and Hanoi. Singapore expresses its distrust of China by urging a continued, strong U.S. naval presence in Southeast Asia. Senior officers of the Singapore Joint Staff and the Singapore Armed Forces Training Institute, May 1999, discussions with author. Indonesia and Thailand are at least more comfortable with that presence than otherwise. Only Malaysia continues to maintain, at least in public, that China is not a threat (and the U.S. naval presence unnecessary). Senior staff officers, Royal Malaysian Defence College in May 1999, discussion with author.

99. Zhang, *FBIS-CHI*-99-0510.

100. Senior Captains Yan Youqiang (Director of a Naval Headquarters Research Institute) and Chen Rongxing, "On Maritime Strategy and the Marine Environment," *Zhongguo Jushi Kexue*, Beijing, No. 2, May 20, 1997, pp. 81-92, in *FBIS-CHI*-97-197. This may be a good description of China's maritime strategic thought.

101. See Alfred D. Wilhelm, Jr., *China and Security in the Asian Pacific Region Through 2010*, Alexandria, VA: Center for Naval Analysis, 1996, p. 44, for the contention that long-range PLAN deployments would be a departure from Chinese military tradition and that PLAN arguments "have not convinced the CMC to allocate the resources" for a large blue-water navy.

102. See, for instance, Marder; Roskill; E.L. Woodward, *Great Britain and the German Navy*, Oxford: Clarendon Press, 1935, and Jonathan Steinberg, *Yesterday's Deterrent: Tirpitz and the Birth of the German Battle Fleet,* London: Macmillan, 1965.

103. This argument is made by Till, p. 186.

ABOUT THE AUTHORS

KENNETH W. ALLEN is a Senior Associate at the Henry L. Stimson Center in Washington, DC, where he directs a project promoting confidence-building measures for China. Prior to joining the Center, he was Executive Vice President of the U.S.-Taiwan Business Council (1992-1998). He served 21 years in the U.S. Air Force (1971-1992), including assignments in Taiwan; Berlin; Headquarters, 5th Air Force in Japan; Headquarters, Pacific Air Forces, China, as the Assistant Air Force Attaché, and the Defense Intelligence Agency. He has written extensively on China's Air Force and China's foreign military relations. He received a B.A. from the University of California at Davis, a B.A. from the University of Maryland in Asian Studies, and an M.A. from Boston University in International Relations.

RICHARD A. BITZINGER is currently a Senior Fellow with the Atlantic Council of the United States, where he is writing a book on the problems and prospects for the global arms industry in the post-Cold War era. Prior to his present position, he worked for the U.S. Government, the RAND Corporation, and the Defense Budget Project, specializing in issues relating to defense industries and the arms trade. He is the author of several monographs, journal articles, and book chapters, including "Military Spending and Foreign Military Acquisitions by the PRC and Taiwan" (in James R. Lilley and Chuck Downs, eds., *Crisis in the Taiwan Strait*, National Defense University Press, 1997), *Gearing Up for High-Tech Warfare? Chinese and Taiwanese Defense Modernization and Implications for Military Confrontation Across the Taiwan Strait* (co-author, February 1996), and "Arms to Go: Chinese Arms Sales to the Third World," *International Security* (Fall 1992).

BERNARD (BUD) D. COLE is Professor of International History at the National War College in Washington, DC. His areas of expertise are Sino-American relations and maritime strategy. He has written one book (*Gunboats and*

Marines: The U.S. Navy in China) and many articles, book reviews, and essays, the most recent of which are "The Chinese People's Liberation Army and Twenty-First Century Warfare," and "China's Maritime Strategy." Dr. Cole earned an A.B. in History from the University of North Carolina, an M.P.A. in National Security Affairs from the University of Washington, and a Ph.D. in History from Auburn University. Dr. Cole previously served 30 years as a surface warfare officer in the Navy. He commanded the USS RATHBURNE (FF1057) and Destroyer Squadron 35. He also served as Naval Gunfire Liaison Officer with the Third Marine Division in Vietnam, Plans and Policies Officer for Commander-in-Chief U.S. Pacific Fleet, and Deputy Director for Expeditionary Warfare on the staff of the Chief of Naval Operations. He has been on the National War College faculty since 1993.

LONNIE HENLEY is Defense Intelligence Officer for East Asia and Pacific, in the Defense Intelligence Agency. He was educated at West Point, Oxford, Columbia, and the Defense Intelligence College. He is a retired Army China Foreign Area Officer, and has worked in a variety of China and Korea-related positions. He is the author of *China's Capacity for Achieving a Revolution in Military Affairs*, U.S. Army War College, Strategic Studies Institute (1996); "The RMA After Next," *Parameters*, Winter 1999-2000; and other publications dealing with Chinese military capabilities, Asian security issues, and the future of warfare.

JAMES R. LILLEY was born in China, educated at Yale University, and served for a number of years in Southeast Asia. He was the Director of the American Institute in Taiwan from 1982-84, the U.S. Ambassador to the People's Republic of China from 1989-91 and to the Republic of Korea from 1986-89. He served as Assistant Secretary of Defense for International Security Affairs from 1991-93 and is presently Resident Fellow at the American Enterprise Institute. Ambassador Lilley is the co-editor of *Beyond MFN: Trade with China and American Interests* (AEI Press, 1994), *Crisis in the Taiwan Strait* (NDU Press, 1997), and *China's Military Faces the Future* (M.E. Sharpe, 1999). He has written extensively on Asian issues.

SUSAN M. PUSKA, a colonel in the U.S. Army, is an Ordnance Officer and China Foreign Area Officer. Colonel Puska has served in logistics assignments in the Republic of Germany; the Republic of Korea; Guantanamo Bay, Cuba; and the United States. She was an Assistant Army Attaché in Beijing between 1992 and 1994, and was the Asia-Pacific Branch Chief and China Desk Officer in the Regional Integration and Assessment Division, Office of the Deputy Under Secretary of the Army, International Affairs between 1996 and 1999. During the 1999-2000 academic year, she served as Director of Asian Studies, Department of National Security and Strategy, U.S. Army War College. She holds a graduate degree from The University of Michigan in Asian Studies, and studied Mandarin Chinese at the Defense Language Institute in Monterey, California, and The Johns Hopkins University–Nanjing University Center for American and Chinese Studies in Nanjing, People's Republic of China. She is the author of *New Century, Old Thinking: The Dangers of the Perceptual Gap in U.S.-China Relations* published by the U.S. Army War College, Strategic Studies Institute, in 1998.

MARK A. STOKES is Country Director for the People's Republic of China (PRC) and Taiwan within the Office of the Secretary of Defense, International Security Affairs (OSD/ISA). He is the author of *China's Strategic Modernization: Implications for the United States* published by the U.S. Army War College, Strategic Studies Institute, in 1999. Major Stokes has served as a signal intelligence and electronic combat support officer in the Philippines and West Berlin. He served as an assistant air attaché at the U.S. Defense Attaché Office in Beijing, PRC, from 1992-95. Before his assignment to OSD/ISA, he was the Asia-Pacific regional planner within the Headquarters U.S. Air Force Operations and Plans Directorate from 1995-97. He holds graduate degrees in International Relations and Asian Studies from Boston University and the Naval

Postgraduate School. He received his formal Chinese Mandarin language training from the Defense Language Institute in Monterey, California, and the Diplomatic Language Services in Rosslyn, Virginia.

WILLIAM C. TRIPLETT II is the former chief Republican counsel to the Senate Foreign Relations Committee. He has 30 years of experience working on China and national security. He is the co-author (with Edward Timperlake) of *Red Dragon Rising: Communist China's Military Threat to America* and *Year of the Rat: How Bill Clinton Compromised U.S. Security for Chinese Cash.*

ARTHUR C. WALDRON is the Lauder Professor of International Relations, China and East Asia, University of Pennsylvania. He is the author of numerous articles on China's military and national security. He is the author of *The Great Wall: From History to Myth* (1990).

INDEX

AA-10a/Alamo, 218
AA-11/Archer, 218
Academy of Military Sciences (AMS), 87, 138
Active defense (*jiji fangyu*), 108, 197, 288–289
AEGIS SPY-1D, 130
Aerospace systems/components, 15
Afghanistan, 133
A-5/Fantans, 215, 222
AGM-154, 265
Airbase blockage (*fengsuo jichang*), 140
Airborne Warning and Control System (AWACS), 23, 130, 146, 265
Airborne laser (ABL), 126
Airbus, 15
Air Combat Maneuvering Instrumentation (ACMI), 206–207
Aircraft industry, civil, 24
Airfield blockages, 144–145
Air power at sea, 306–307
Air superiority (*zhikongquan*), 110, 140
Air tasking order (ATO), 229
Air-to-air missiles (AAMs), 14, 19, 21, 218
Albright, Madeleine, 256, 266

Aleutians, 293
All-Army Joint Logistics Implementation Plan, 56
Allen, Ken, 20
Alliance of Democratic Countries, 259
Allied Signal, 15
AL-31F, 22
American Enterprise Institute, 2
Amur River, 290
AN-12/Cub, 221
AN-24/Coke, 219
AN-26/Curl, 219
Antiaircraft artillery (AAA), 191
Anti-Americanism, 256, 272
Anti-radiation missiles, 130
Antiship cruise missiles (ASCMs), 11, 18–19
AN/TPQ-37 radar, 23
Arayama, Yuko, 32, 33, 39
Argus radar system, 220
Arms production, technology leveraging and, 24–28
Ascent anti-satellite (ASAT) program, 139
Aspide AAM/SAM, 14, 21
AS-350 helicopters, 14
AS-365N helicopters, 20
AT-4 *Spigot*, 21
ATR, 15

Attack opportunities/sites, 144–145

Attack Regiment, 13th, 203

Averting an Electronic Waterloo, 89

Aviation Industries of China (AVIC), 34, 35, 36, 37

Aviation industry
 commercial, 15
 debate on importing versus producing weapon systems, 194

Aviation troops/branch (*hangkongbing*), 202–205

AWACS. *See* Airborne Warning and Control System

B-2A Spirit, 265

B-5/Beagle, 215, 218, 222

B-6/Badger, 192, 218, 222

Ballistic missiles, theater, 119–127

Battle damage assessment (BDA), 112

Beihai, 298

Beijing Military Region, 63, 202

Beijing University, 85

Belarus, 118

Beriev, 220

BGC-161, 117

Blue Army aggressor units, 208–211

Blue-Water navy, 279–280, 300–307

Boeing Aircraft Co., 15, 265

Bombardier, 15

Bomber Regiment, 12th, 203

Bombers, 218–219

Bonin Islands, 280, 294

Brazil, 13, 115, 118

British Aerospace, 36

C-101 ASCMs, 21

C-301 ASCMs, 21

C-601 ASCMs, 222

C-701 ASCMs, 19

C-801 ASCMs, 18, 222

C-802 ASCMs, 11, 18, 21

Campaign mobility phase (*zhanyi jidong jieduan*), 142–143

Canada, assistance from, 13, 15, 114

Cangxian, Flight Test and Training Center at, 208–209

Cao Gangchuan, 39

Caroline Islands, 280, 294

Cao Shuangming, 193

CBU-78, 123

CCK (Chingchuangkang) Airbase, Taiwan, 83, 145

Center for Strategic and International Studies (CSIS), 88–89, 98

Central Military Commission (CMC), 56, 109, 112, 118, 189, 200, 230

CFM-56 turbofan engines, 14

334

Charged couple devices (CCDs), 115
Chengdu Aircraft Co., 194
Chengdu Military Region, 202, 216
Cheng Mingshang, 296
Chengshan, 297
Chiang Kai-shek, 285
Chile, 139
China Academy of Launch Technology (CALT), 119, 123, 125
China Aerospace Corp. (CASC), 34, 35, 36
China Aerospace S&T Corp. (CASC), 113, 115, 122, 126, 127, 129, 133, 138
   Sanjiang Space Corp. (066 Base), 119, 120
China-Brazil Earth Reconnaissance Satellite (CBERS), 13, 22
China Democratic Party, 275
China Economic Times, 257
China Launch and Tracking Control General (CLTC), 113, 139
China Military Science, 86
China National Nuclear Corp. (CNNC), 34, 35, 36
China Ordinance Industry Corp. (COIC/ NORINCO), 34, 35, 36
China State Shipbuilding Corp. (CSSC), 34, 35, 36, 37
China Times Express, 212
Chinese Communist Party (CCP), 260
"Chinese People's Liberation Army Joint Campaign Programs," 69
Chinese PLA Military Terminology, 71
Chingchuankang (CCK) airfield, 83, 145
Choke points, 300
Circular error of probability (CEP), 120, 128
Clausewitz, 267
Clinton, Bill, 98, 263, 264
Combat regulations, 69–70, 71
Combat Regulations Compilation Committee, 69
Combined Brigade, 4th, 204
Command, control, communications, computer, and intelligence (C^4I), 11, 143, 145
   improved interconnectivity, 12
Command and control (C^2), 143
Commercial off-the-shelf (COTS) technologies, 27
Commercial technology transfers, 14–17
Commission on Science, Technology, and Industry for National Defense

335

(COSTIND), 12, 33–34, 39, 82, 87, 113–114, 138
"Common and Garrison Regulations," 72
Common Regulations, 71
Communications Command Academy, 87
Communications command vehicle (*tongxun zhihuiche*), 137
Communications regiment (*tongxintuan*), 136
Communications security (COMSEC), 137–138
Computer center (*jisuan zhongxin*), 136
Computer network attack (CAN), 79
U.S. vulnerability to, 88–94
Computers, high performance, 83–84
Conventional-takeoff-and-landing (CTOL), 306
Corbett, Julian, 280
CO laser guidance system, 135
Counterattack phase (*kangki fanji jieduan*), 146–147
Counter-reconnaissance (*fanzhencha*), 146
Counterspace measures, 138–139
Cox-Dicks Committee, 83–84, 85
Cox Report, 14, 15, 29
Critical civilian infrastructure, 79

Critical infrastructure, U.S., 89–90
Crotale short-range SAM, 14, 19
Cruise missiles
See also Theater missiles; *under type of*
advancements in, 18–209
conclusions, 40, 41
imported technology, 12–13
CSS-N-3 IRBMs, 306
Cuba, 83
Danwei, 31
Data relay satellite (DRS), 117
Da Xinganling, 199
Deep attack strategy (*zongshen daji*), 119
Defense industry
conclusions, 38–41
cultural impediments, 31–33
enterprise groups, 34–38
rationalization, 37
reforming the, 33–38
reorganization, 33–34
restructuring, 34–37
structural impediments, 30–31
technical impediments, 28–29
Defense Science Board, 88, 90–91
Deng Xiaoping, 61, 260, 261, 264, 279, 292, 307
Depressed trajectories, 126–127
Desert Storm, 303, 305
Deutch, John, 88
DF-3 missile system, 127

336

DF-11 missile system, 18, 120, 123
DF-15 missile system, 18, 119–120, 124
DF-21 missile system, 116, 120–121, 126, 144
DF-25 missile system, 160–161
DF-31 missile system, 116
Diaoyutai, 296
Dictionary of Military and Associated Terms (Joint Publication 1-02), 71
Digital scene matching area correlation (DSMAC) system, 132
Dinghai, 298
Ding Henggao, 39
Dingxin, 211
Distribute Control Systems, 93
EA-6B Prowler, 265
East China Sea, 279, 293
East Timor, 275
ECM regiment (*dianzi duikangtuan*), 136
80302 Unit, 136
863 Program, 12, 113, 118
Eight Points of Attention, 60
Eisenhower, Dwight, 287
Electric power generation vehicle (*fadianche*), 137
Electromagnetic pulse warhead (EMP), 123
Electronic countermeasures, 138

Electronic identification, 231–232
Electronic intelligence (ELINT), 13, 220, 221
Electronic reconnaissance satellites, 113, 114–115
Electronic surveillance, 220
Electronic warfare, 80, 220–221
Electro-optical systems, 114, 115–116
Elta, 220
Encryption technology (*mimaxue*), 138
Enemy counterattack phase (*kangki fanji jieduan*), 146–147
Engineering Institute, 87
Engineering regiment (*gongchengtuan*), 136
Equipment assurance units (*zhuangbei baozhang budui*), 136
subunits, 137
ERS-1 satellite, 114
Ethnic cleansing, 257, 260, 261
Eurocopter, 15
Exocet, 18, 21
EY-8, 221
F-10 fighters, 13
F-16 fighters, 192
F-117A fighters, 146–147
Falun Gong, 275
Fanhuishi-2 (FSW-2), 116
FB-7, 221, 222, 232

FC-1 fighters, 13, 23
Fighter aircraft, 13, 20, 23–24
Film-based recoverable satellites (FSW-3), 116
Financial markets, effects of war on, 270
Financing defense market, 30–31
Firepower coordination cell (*huoli xietiaozu*), 141
Firing battalions (*fasheying*), 137
Fisher, Richard, 12, 13
Five Year Plan, 114
Fleet Ballistic Missile (FBM) submarines, 291
Flight Test and Training Center at Cangxian, 208–209
FM-80 SAM, 14, 19
FM-90 SAM, 19
Food service reforms, 66–67
Foreign intervention, preventing, 111–112
France, 14, 18, 118, 139, 192
high-powered microwave weapons (HPMs), 85
Frieman, Wendy, 18, 26
FSW-2, 116
FSW-3, 116
FT-2000 mobile SAM, 224
Fujian MD, 67, 225, 230, 268
Fu Quanyou, 82
Fuzhou, 224

Gaining the initiative by striking first (*xianfa zhiren*), 110
Gang of Four, 290–291
Gansu province, 83, 145
Gobi Desert training center, 211–213
Jiuquan Space Launch Center, 116
GEC-Marconi, 220
General Armaments Department (GAD), 34, 39, 113, 134, 138, 139, 195
General Logistics Department (GLD), 55, 56–57
Wuhan Rear Base, 57
General Political Department, 82
General Staff Department (GSD), 71, 87, 131
Geographic information systems (GIS), 131
Germany, joint ventures with, 16, 118
Gill, Bates, 12
Global Navigation Satellite System (GLONASS), 18, 131
Global positioning system (GPS) satellites, 11, 18, 122, 146
NAVSTAR, 131, 138
Gobi Desert training center, 211–213
Godwin, Paul, 197
Golf class submarine, 305–306

Gorshkov, Sergei, 291, 295
Gray, Colin, 281
Greater Khingan Mountains, 199
Great Proletarian Cultural Revolution (GPCR), 290
Ground control intercept (GCI), 219
Ground processing, 116–118
Guangdong province, 201
Guangxi, 199
Guangzhou Military Region, 63, 202, 216, 298
  upgrading MRAF, 213–214
Guangzhou Yangcheng Wanbao, 213
Gulf War, 69, 109, 148, 266
Gunboat policy, 260, 282
Guningtou battle, 268
Haikou, 298
Hainan, trial programs in, 57
Haitan Island, 199
Haiyang-1 (HY-1), 114
Han class submarine, 22
Hangzhou, 298
Harbin Institute of Technology, 130
Hart, Gary, 90
Hexcell, 15
High performance computers (HPCs), 83–84
High-powered microwave weapons (HPMs), 85, 123

HJ-8, 21
HN-1 LACMs, 19
Hong Kong, 270
HQ-9 medium-range SAM program, 13
Huahong Co., 16–17, 22
Huangpi, 198
Huangpu, 298
Huangshan base, 136
Hubei province, 67
  Sanjiang Space Corp. (066 Base), 119, 120
Huludao, 297
Human rights, 257, 260, 261
HY-4, 129
Identification friend or foe (IFF) system, 231
IL-14 transports, 203
IL-28/Beagles, 218
IL-76 transports, 192, 200, 214, 219, 220
Imported technology, 12–14
India, 197, 218, 273, 289
  China's maritime strategy and, 299
Indian Ocean, 112, 279, 299
Indonesia, 274–275, 293
Inertial navigation system (INS), 210
Information dominance (*zhixinxiquan*), 110, 111, 140
Information Engineering University (Zhenzhou), 83

Information Warfare (IW)
  applications of PLA, on the Untied States, 88–94
  definitions and scope, 80–81
  financial resources, 82–83
  summary of PLA capabilities and doctrine, 81–88
  United States, 79–80
  United States response to, 98–99

Infrared signature reduction, 134

Institute of Applied Physics, 85

Integrated Air Defense System (IADS), 223

Intelligence, surveillance, and reconnaissance (ISR), 113

Intercontinental cruise ballistic missiles (ICBMs), 18

Intermediate Range Ballistic Missiles (IRBMs), 306

Iraq, 271

Israel, 271
  assistance from, 12, 13, 14, 23, 130, 192, 194, 216, 221

Israeli Aircraft Industries (IAI), 220

Italy, 14, 117, 118

J-6 fighters, 200, 215

J-7 fighters, 20, 21, 190, 200, 210, 215

J-8 fighters, 190, 200, 210, 215

J-8IIM fighters, 13, 20, 21, 23, 192

J-10 fighters, 13, 20, 21, 22, 190, 215, 216, 217

J-11 fighters, 190, 215, 216, 217

Jane's Defence Weekly, 19, 198, 221

Japan
  joint ventures with, 16
  relations with, 262, 263–264, 274, 293, 297
  vulnerability to Information Warfare, 96–97

Japan Commission on Critical Infrastructure Protection, 96

Japan National Railways (JNR), 81

JERS-1 satellite, 114

JH-7 fighters, 20, 23

Jiang Jiesheng, 57–58

Jiang Qing, 290

Jiangxi province, 136

Jiang Zemin, 61, 62, 68, 84, 190, 198, 260, 264, 272, 279, 282, 296

Jiayuguan, 145

Jiefangjun Bao, 57–58, 60–61, 189, 208, 209, 211–212, 223–224

Jinan, trial programs in, 57

Jinan Military Region, 71, 202, 297

Jinmen, 227, 268

Joint Air-to-Surface Standoff Missile (JASSM), 265

Joint Direct Attack Munitions (JDAM), 265
Joint Logistics Departments (JLDs), 58
Joint Publication 1-02 (Dictionary of Military and Associated Terms), 71
Joint Publication 3-01.5, 151
Joint service logistics, 57–58
Joint Standoff Weapon (JDAM), 265
Joint Surveillance Target Attack Radar System (J-STARS), 146, 265
Kaifeng, 198
Kaohsiung, 228
Kazakhstan, 272
Kh-31P, 130
Kh-55/AS-15 strategic cruise missile, 19
Kh-55 strategic cruise missile, 19
Kim, Taeho, 12
Kiribati, 139
Korean War, 190, 191, 197, 200, 288–289
Kosovo War, 229
  Allies/NATO, role of, 265–267
  Chinese responses to, 271–274
  force, use of, 261–265
  implications of, for China, 244–257
  implications of, for Taiwan, 260–261
  international aspects, 270–271
  precision guided munitions, role of, 267–268
  public relations aspects, 269–270
  sovereignty, international law of, 257–260
  theater missile defense, 268–269
Krushchev, Nikita, 289
Kunlun Plateau, 199
Kunming, 139
Kuomintang (KMT), 285
Kuriles, 293, 294
Land-attack cruise missiles (LACMs), 11, 19
  anti-radiation, 130
  guidance systems, 135
  infrared signature reduction, 134
  mission planning, 131–133
  production trends, 133
  programs, 129
  propulsion systems, 133–134
  radar signature reduction, 134
  reasons for, 128
  role of, 127
  Silkworm, 129
  YJ-8, 129–130
  YJ-9, 129, 130
Lanzhou Military Region, 202
Information Warfare, 88
Large-scale integrated circuits (LSICs), 121
Laser cladding, 126

341

Launch sites, pre-surveyed and reserved, 137
Lavi fighter jet program, 13, 216
Law of the Territorial Sea and Contiguous Zones, 298
Lebedev Physics Institute, 85
Lee Teng-hui, 192, 213, 225, 261, 270
Leveraging technology, assessment of, 17–28
Lewis, John, 205–206
Liberation Army Daily (LAD), 81, 82, 87
Lin Biao, 289–290
Lin Chong-pin, 95, 96
Lin Weigan, 85
Liu Huaqing, 82, 280, 293–295, 298, 300, 301, 307
Liu Shunyao, 189, 191, 192, 193, 197–198, 199, 205, 207, 208, 224, 228, 235
Lockheed Martin, 36, 265
Logistics reform, PLA, 55
 food service reforms, 66–67
 funding for, 66
 joint service logistics, 57–58
 military regions, role of, 59
 mobile logistics, 62–64
 monetization of officer compensation system, 64–66
 objectives of, 56
 other changes, 66–67
 skip-echelon logistics, 66
 socialized logistics, 59–62
 standardization of supply procedures, 65
 trial programs, 56–57
Low observability (LO) technological capabilities, 11
Luda destroyers, 292
Luhai destroyers, 22, 26
Lu Linzhi, 87
Lushun, 297
Luzon Strait 300
LY-60 SAM, 14
M-9 missile system, 18
M-11 missile system, 18, 120
Mao Zedong, 196, 197, 264, 279, 286, 288, 289, 290
Mariana Islands, 280, 294
Maritime strategy
 background information, 279–282
 Blue-Water navy, 279–280, 300–307
 conclusions, 312–314
 evolution of, 284–295
 factors influencing, 281
 India, 299
 Liu Huaqing's vision, 293–295
 modernization of navy, 283–284
 in northeast Asia, 297
 politics, geography and, 282–283
 prospects, 307–311
 sea lines of communications (SLOCs), 291, 299–300

South China Sea, 298–299
  strategic interests, 296–300
  Taiwan and, 297–298
Matsu, 230, 288
Mazu, 227
McDonnell Douglas Corp., 15
MD-80 and MD-90 passenger jets, 15
Medeiros, Evan, 15–16, 27
Microelectonics, 16–17, 22
Mid-course guidance (*zhongduan zhidao*), 146
MiG-21, 191
MiG-23, 23
Military region air force (MRAF), 201
Military Regions (MR), 57
  mobility and, 63
  role of, 59, 202
Millimeter wave seeker (MMW), 121–122
Milosevic, Slobodan, 255, 256, 259–260, 266
Ming class submarine, 20, 22
Ming dynasty, 284–285
Ministry of State Security, 81
Mirage 2000-5s, 192, 226, 227
Mischief Reef, 299
Missiles. *See under type of*
Missile strike phase (*daodan tuji jieduan*), 143–146
Missile transport vehicle (*daodan yunshuche*), 137

Missile/warhead storage unit (*zhuangbei jishu qinwu budui*), 136
Mission planning, 131–133
Miyawaki, Raisuke, 96
Mobile logistics, 62–64
Molander, Roger, 79–80
Mongolia, 264
Mortal blow (*zhiming daji*), 110
Mourdoukoutas, Panos, 32, 33, 39
Mulvenon, James, 24, 25, 28–29, 31
Nanjing Military Region, 57, 139, 141, 202, 203, 216, 228, 230, 298
National Bureau of Oceanography, 201
National Counterintelligence Center, 84
National Defense University, 87, 108
Nationalist Air Force, 191
National People's Congress (NPC), Ninth, 33, 298
National University of Defense Technology, 125–126
NATO (North Atlantic Treaty Organization), 229, 255, 259, 263, 265–268
Naval air force, 221–223
Naval War College (U.S.), 264
NAVSTAR GPS, 131, 138
NEC Corp., 16–17, 22

New High Ground of Strategic Competition, 310

New York Times, 261

909 Project, 16–17, 22

Ningbo, 298

North Industries Corp. (NORINCO), 34, 35, 36–37

North Korea, 271, 273, 297

Ocean monitoring/surveillance (*haiyang jianshi*), 113, 146

Officer compensation system, 64–66

Okean exercises, 292

Okinawa, 134

Olivetti image processing, 117

Operational preparations phase (*zuozhan zhunbei jieduan*), 141–142

Operational test and evaluation (OT&E), 222

Overcapacity problems, 30

Pacific Ocean, 112, 279

Pakistan, 271

Paracel Islands, 292, 298

Patriot SAM technology, 13, 124, 125, 130

Payloads, 122

Peng Dehuai, 289

People's Daily, 256

People's Liberation Army (PLA)
See also Logistics reform, PLA
assessment of reforms, 72–74
Combat Order No. 13, 56
combat regulations/operational ordinance, new, 55
debate on the modernization of, 1–2, 9
logistics reform, 55, 56–67
1999 Conference, 2
reorganization, 72
standardization, 68–71
summary of Information Warfare capabilities and doctrine, 81–88

People's Liberation Army Air Force (PLAAF), 62, 63–64, 140
activity over the Taiwan Strait, case study, 225–232
airborne early warning, 220
airborne troops, 198–199
air defense troops, 223–224
air-to-air missiles, 218
aviation branch, 202–205
Blue Army aggressor units, 208–211
bombers, 218–219
Brigade (43˚), 198
Brigade (44˚), 198
Brigade (45˚), 198
Command College, 195
commander's assessment, 192–196
doctrine and strategy, 196–198
15˚ Airborne Army, 192, 198–199
fighters, 215–218
flight training, 205–223
Gobi Desert training center, 211–213

344

intelligence collection aircraft, 220–221

mission and organization of, 200–205

naval, 221–223

rapid-reaction force, 200

transports, 219

upgrading of, 189–192

upgrading the Guangzhou MRAF, 213–214

weapons systems, 214–215

People's Liberation Army Navy (PLAN), 63

See also Maritime strategy

air power at sea, 306–307

capabilities, 302

doctrine, 302–305

East Sea Fleet, role of, 298

formation of, 286

Liu Huaqing's vision, 293–295

modernization of, 283–284, 301

North Sea Fleet, role of, 297

nuclear deterrence and, 305–307

prospects, 307–311

South Sea Fleet, role of, 298–299

People's Navy, 222

People's war doctrine, 196, 288, 290

Pershing-II, 121, 125

Phalcon AEW radar, 220

Philippines, 293, 299

Photoreconnaissance, 12

PL-7 AAM, 19

PL-9 AAM, 14, 19

PL-11 AAM, 14, 21

Pollack, Jonathan, 24, 25, 28–29, 31

Pratt & Whitney, 15

Precision-guided munitions/missiles (PGMs), 14, 192, 267–268

Precision strikes, 85–86

Preemptive strikes, 87, 109, 150

President Decision Directive (PDD) 63 "Critical Infrastructure Protection," 98

President's Commission on Critical Infrastructure Protection, 89–90, 92–94, 95

Product level, technology leveraging at the, 17–24

Project-094 boat, 306

"Provisional Stipulations for Grading Military Training," 71

Pursuit Regiments, 10th and 11th, 203

Pyongyang, 273

Python-3 AAM, 14

Qian Xuesen, 82

Qing dynasty, 285

Qinghua University, 85

Quality control issues, 26

Quemoy, 230, 288

Quingdao, 67, 297

QW-1 SAM, 19
Radar
 -absorbent materials, 11, 134
 imaging satellites, 113–114
 imported, 13
 signature reduction, 134
 solid-state phased-array, 12
 ultra-wide band, 12
RADARSAT, 114
Radio frequency weapons, 85
RAND, 86, 89, 97, 98
Rapid reaction (*kuaisu fanying*), 142
Rapid-reaction force (RRF), 198–200
Rapid war, rapid resolution (*suzhan, sujue*), 110, 148
Raytheon, 265
Reconnaissance satellites, 13, 22
Reconaissance unit (*jizhen dadui*), 136
Red Army, 209
Reentry vehicles, 124–125
Refineries, vulnerability of U.S., 93–94
Remote sensing, 12
Reorganization, PLA, 72
Repair depot (*tezhuang xiulicang*), 136
Republic of China: 1998 National Defense Report, 228–229
Research and development (R&D)

Information Warfare centers, 83
 plans, 12
Reversed-engineered military equipment, 14, 23–24
Revolution in Military Affairs (RMA), 79, 109, 267
R550 *Magic* AAM, 19
RSBN (radio navigation aid), 210
Rubin, 14
Rudman, Warren, 90
Rugao airfield, 215
Rules-of-engagement (ROEs), 230
Russia, 130, 263, 272
 assistance from, 12, 13, 14, 18, 19, 122, 192, 194, 216–218, 220, 224, 306
 cooperative efforts, 118
 high-powered microwave weapons (HPMs), 85
Ryukyus, 293
SA-10 SAMs, 13, 224
SA-15 SAMs, 23, 224
SA-321G Super *Frelon* helicopters, 23
Sanjiang Space Corp. (066 Base), 119, 120
SATCOM, 143
Science and Engineering University, 83
Science and Technology Daily, 194

346

Science and technology (S&T) plans, 12
Scramjet (supersonic combustion ramjet engine), 134
SCUD-B, 121
Sea lines of communications (SLOCs), 291, 299–300
Sea superiority (*zhihaiquan*), 140, 141
Second Artillery, 126
　See also Theater missile campaign
　information denial, 137–139
　organization of, 135–137
Seek Optics, 11
Self-Strengthening Movement, 282
Senkaku Islands, 296
Serbia. *See* Kosovo War
Shanghai, 270, 298
Shanghai Academy of Spaceflight Technology (SAST), 115
Shanghai Bureau of Astronautics' 701 Program, 115
Shantou, 298
Shenyang Aircraft Co., 194
Shenyang Military Region, 202, 297
　Information Warfare, 88
　trial programs in, 57
Shichang, 67

Shipbuilding industry, 15–16, 19–20, 25, 26, 27
Shiwan Mountains, 199
Shi Yunsheng, 302, 307–308
Signals intelligence (SIGINT), 83, 220
Sikorsky Helicopter, 15
Silkworm, 18, 129
Simulators, 207–208
SkyMaster AEW, 222–223
Soldiers' Service Regulation, 66
Socialized logistics, 59–62
Song class submarine, 19–20, 21, 22
South China Sea, 112, 262, 279, 280, 292, 293
　China's maritime strategy in, 298–299
South Korea, 16, 262, 263, 271, 274
South Vietnam, 292
Sovereignty, international law of, 257–260
Soviet Union
　assistance from, 286–287
　China's split with, 289, 290
　collapse of, 262
　Navy in the 1960s and 1970s, 291–292
Sovremenny destroyers, 23, 307
Space support, for theater missiles, 112–118
Space tracking network, 139
Special Operations, 86

Speys, 22, 23
Spratly Islands, 298, 306
SSBN nuclear ballistic-missile-carrying submarines, 14
SSK diesel submarine, 19–20
SSN nuclear attack submarines, 14
SS-N-2 *Styx,* 18
Standardization, PLA, 68–71
STAR-1 ARM system, 130
State-owned enterprises (SOS), 29, 30–31, 32
State Science and Technology Commission (SSTC), 113–114
S-300, 191, 192, 224
Stinger, 19
Strait of Hormuz, 300
Strait of Malacca, 300
Strategic Information Warfare—A New Face of War (RAND), 86, 89
Stokes, Mark, 29
Submarine-launched ballistic missiles (SLBMs), 18
Submarines
  diesel, 19–20, 21, 22
  Golf class, 305–306
  Han class, 22
  imported, 14
  Ming class submarine, 20, 22
  nuclear-powered attack, 292
  Song class, 19–20, 21, 22

Xia class, 306
Submunitions, 123
Suixi, 225
Super-863 Program, 12
Supervisory Control and Data Acquisition (SCADA), 92–93
Surface-to-air missiles (SAMs), 13, 19, 191
Surface-to-surface missiles (SSMs), 18, 144–145
Surveying/mapping unit (*cehui dadui*), 136
Surveying vehicle (*cekongche*), 137
SU-27 fighters, 13, 22, 23, 24–25, 190, 191, 192, 200, 210, 215, 216, 217
SU-27SK, 217
SU-27UBK, 217
SU-30 fighters, 23, 190, 191, 215
SU-30MKI, 218
SU-30MKK, 217, 218
Synthetic aperture radar (SAR), 11, 13, 113–114, 135
Systems integration capabilities, 28–29
Table of organization and equipment (TO&E), 204
Tactical Air Defense System (TADS), 223
Tactical Air Navigation (TACAN), 210, 232
Tactical imaging system, 115–116

Taichung, 145

Taipei, 140, 151, 228

two states theory, 261

vulnerability to Information Warfare, 95–96

Taiwan, 70, 82, 136, 140, 288, 293
  CCK (Chingchuangkang) Airbase, Taiwan, 83, 212
  China's maritime strategy and, 297–298
  implications for, 148–153
  Kosovo War, implications of, 260–261, 262
  as a primary force, 110–111
  simulated attacks on, 145, 212
  vulnerability to Information Warfare, 94–96, 97

Taiwan Air Force, 140, 145, 227, 228, 230

Taiwan Ministry of National Defense (MND), 225, 227

Taiwan Relations Act, 99

Taiwan Strait, 88, 94, 149, 300
  Crisis of 1958, 190, 200, 227
  military exercises in, 199, 224, 225–232

Tang Fei, 94

Tang Yaoming, 95–96

Tarim Desert Highway, 67

Technical position (*jishu zhendi*), 136

Technology
  See also under type of
  acquisition activities, 11–17
  acquisition versus leverage of, 9
  commercial transfers, 14–17
  conclusions, 38–41
  cultural impediments, 31–33
  impediments to absorption and exploitation, 28–33
  imported, 12–14
  leveraging, assessment of, 17–28
  program stretchouts and small production runs, 20–21
  reversed-engineered, 14, 23–24
  structural impediments, 30–31
  technical deficiencies, 28–29

Telecommunications infrastructure, 83

Telespazio, 117

Terminal guidance, 146
  global positioning system, 122
  millimeter wave seeker, 121–122
  terrain matching, 121

Terrain contour matching guidance (TERCOM), 11, 131–132

Thailand, 306

Theater command center (*zhanyi zuozhan zhongxin*), 141

Theater missile campaign
  conclusions, 147–153
  countermeasures, 151–153
  counterstrike phase, 146–147
  mobility phase, 142–143

operational implications, 148–149

operational preparations phase, 141–142

phased campaign, 139–141

political implications, 149–151

strike phase, 143–146

Theater missile defense (TMD), 282–283

limitations of, 268–269

Theater missiles

countermeasures, 124–127

developments, 118–135

electronic reconnaissance satellites, 114–115

electro-optical reconnaissance systems, 115–116

film-based recoverable satellites (FSW-3), 116

ground processing, 116–118

land attack cruise missiles, 127–135

radar imaging satellites, 113–114

reasons for Beijing's reliance on, 109–112

role of, 107–108

space support for, 112–118

Theater missiles, ballistic

countermeasures, 124–127

depressed trajectories, 126–127

DF-3 missile system, 127

DF-11 missile system, 18, 120, 123

DF-15 missile system, 18, 119–120, 124

DF-21 missile system, 116, 120–121, 126

DF-25 missile system, 160–161

DF-31 missile system, 116

laser cladding, 126

payloads, 122

production rates, 119

submunitions, 123

terminal guidance, 121–122

warheads, 123–124

Third Academy, 127, 129, 130, 132–135

Three Gorges Dam, 258

Tiananmen, 219, 261, 267

Tianjin, 227

Tibet, 199, 260, 273

Tomahawk missiles, 133, 256, 266

Tor-M1 regiment, 224

Toshiba Machine, 98

Training

flight, 205–223

for Information Warfare, 87–88

standards, 71

Transfer point (*zhuanzai changping*), 136, 142

Transfer station (*zhuanyunzhan*), 136

Transports, 219

Truman, Harry, 287

Trunkliner program, 15

TU-4, 220

TU-16/Badgers, 218

Turkey, 272
Two states theory, 261
Type-704, 23
Ukraine, 118
United Kingdom
  assistance from, 222
  high-powered microwave weapons (HPMs), 85
United Nations, 259
United States
  dual-use information technologies form, 17
  high-powered microwave weapons (HPMs), 85
  vulnerability to computer network attack, 88–94, 9899
U.S. Air Force, 112
U.S. Army, 265
U.S. Army Field Manual 100-6, 86
U.S. Army War College, Strategic Studies Institute, 2
U.S. Commission on National Security/21st Century, 90
U.S. Department of Defense, 151, 199, 204, 215, 217, 218–219, 221–222, 223
U.S. General Accounting Office (GAO), 88, 98
U.S. Marine Corps, 265
U.S. Navy, 265
U.S. Office of Naval Intelligence, 226, 229
U.S. Seventh Fleet, 287
U.S. Strategic Air Command, 83
Unmanned Aerial Vehicles (UAV), 85
Very large-scale integrated circuits (VLSICs), 121
Vietnam, 213, 261, 274, 289
Vietnam War, 151–152, 190, 197, 266
Visual flight regulations (VFR), 193
Vital points (*dianxue*), 110
Waigaoqiao ship yard, 30
Wang Baocun, 95
Wang Baoqing, 99
Wang Ke, 55, 56, 58, 59, 61, 64, 65
Wang Xiaodong, 274
Wang Yizhou, 262
Warheads
  CBU-78 GATOR-like, 123
  decoys, 125–126
  electromagnetic pulse (EMP), 123
  fuel-air explosive (FAE), 124
  high-powered microwave (HPM), 123
  penetration, 123–124
  shaping, 125
Weather center (*qixiang zhongxin*), 136
Weihai, 297
Wen Wei Pao, 86

Western Europe, imported technology from, 12
Wide Angle, 59, 60, 65
Winning victory with one strike (*yizhan, ersheng*), 110
Workforce
  downsizing, 34
  problems, 29–31
World Economics and Politics Research Institute, 262
World Trade Organization (WTO), 257
World War I, 286
World War II, 266
Wortzel, Larry, 23, 32–33
Wu Guangyu, 190, 208
Wuhu, 225
Wusong, 298
X-600 LACMs, 19
Xia class submarine, 306
Xiamen, 67, 298
Xian Aircraft Co., 194
Xiangshan, 298
Xiao Jingguang, 287
Xinhua, 59, 60, 61, 201, 211, 213
Xinjiang MD, 67, 199, 260, 263, 272, 275
Xue Litai, 205–206
XY-41 missile, 129
Y-5/Colt, 219
Y-8 transports, 214, 219
Yalu River, 297

Yan'an Spirit, 60
Yangtze River, 202
Yellow Sea, 279, 293
Yingji-8 missile (YJ-8), 129–130
Yingji-9 missile (YJ-9), 129, 130
Yingshan, 198
Young School, 286, 291
Yuchi, 297
Yugoslavia, 148
  See also Kosovo War
Yulin, 298
Yun-5 transports, 203
Z-8 helicopters, 23
Z-9 helicopters, 20
Z-11 helicopters, 14
Zhang Chunqiao, 291
Zhang Wannian, 61–62, 283
Zhanjiang, 298
Zheng He, 284
Zhu Bangzao, 257
Zhu Guang, 205
Zhujiang River delta area, 201
Ziyuan-1 (ZY-1), 115